ELASTIC SCATTERING OF ELECTROMAGNETIC RADIATION

Analytic Solutions in Diverse Backgrounds

ELASTIC SCATTERING OF ELECTROMAGNETIC RADIATION

Analytic Solutions in Diverse Backgrounds

SUBODH KUMAR SHARMA

CRC Press
Taylor & Francis Group
Boca Raton London New York

CRC Press is an imprint of the
Taylor & Francis Group, an **informa** business

CRC Press
Taylor & Francis Group
6000 Broken Sound Parkway NW, Suite 300
Boca Raton, FL 33487-2742

Printed on acid-free paper
Version Date: 20171221

International Standard Book Number-13: 978-1-4987-4857-5 (Hardback)

Visit the Taylor & Francis Web site at
http://www.taylorandfrancis.com

and the CRC Press Web site at
http://www.crcpress.com

Dedicated to fond memories of my parents
and to my dear
Shibani, Kunal, Lori, and Aahana

Contents

Preface

The technique of elastic scattering of electromagnetic radiation is used extensively for characterizing single particles as well as particulate systems. This technique is a truly multidisciplinary technique, and has found applications in diverse disciplines of science, engineering, medicine, agriculture, etc.

Many books exist on this topic. However, most of them focus on specific themes. Some, restrict the scope by choosing a particular spectral region of the electromagnetic spectrum. Consequently, there appears to be room for a book with a wider perspective. This monograph is a modest attempt to fill this space. An effort has been made to keep the presentation self-contained and simple. It is hoped that this monograph will help the reader to become aware of developments across the disciplines, and provide a flavour of the basic analytic approaches.

The subject matter is limited to analytic solutions. These are mathematical expressions constructed using known mathematical operations. As a result, the analytic solutions are mostly approximate. From the point of view of the target, the stress is on single particle scattering and on particulate systems in which multiple scatterings are negligible. Some peripheral aspects of the multiple scattering problem are included.

Many friends and colleagues have helped in the preparation of this monograph, in different ways. I would like to thank them all. In particular, I would like to mention the names of Dr. Ratan K. Saha and Mr. Ransell R. Dsouza for helping me with preparation of a number of figures. Last but not the least, I would also like to thank the publishers and printers for all the help provided by them. It was a pleasure to work with them.

S. K. Sharma
Kolkata, India

List of Figures

List of Tables

Symbols

ADA — anomalous diffraction approximation.
ADr — anomalous diffraction rapid.
ADT — anomalous diffraction theory.
α_p — polarizability.
$\mathbf{B}$ — magnetic induction vector.
BNA — Bohren and Nevitt approximation.
$\langle\cos\theta\rangle$ — first moment of the phase function.
C_{abs} — absorption cross section.
C_{sca} — scattering cross section.
C_{ext} — extinction cross section.
$CASPF$ — Cornette and Shanks phase function.
CLD — chord length distribution.
CSA — Caldas and Semião approximation.
$CSPF$ — Caldas and Semião phase function.
$CSPFL$ — Caldas and Semião phase function, large particle.
$\mathbf{D}$ — dielectric displacement.
DDA — discrete dipole approximation.
DPF — Draine phase function.
$\mathbf{E}$ — electric field vector.
ε — dielectric constant.
ε_r — real part of dielectric constant.
ε_i — imaginary part of dielectric constant.
EA — eikonal approximation.
$EADA$ — extended ADA.
EFA — Evans and Fournier approximation.
$FFPF$ — Fournier and Forand phase function.
$FPPF$ — five-parameter phase function.
FMA — Fymat–Mease approximation.
g — asymmetry parameter.
GEA — generalized eikonal approximation.
$GGADT$ — general geometry ADT.
$GKPF$ — Gegenbauer kernel phase function.
GRA — Gaussian ray approximation.
$GLMT$ — generalized Lorenz–Mie theory.
$\mathbf{H}$ — magnetic field vector.
$h_n^{(1)}$ — Hankel function of first kind.
HMA — Hart and Montroll approximation.
$HGPF$ — Henyey–Greenstein phase function.
$h_n^{(2)}$ — Hankel function of second kind.
I_{inc} — incident intensity.
I_{sca} — scattered intensity.
$i_\perp(\theta)$ — $|S_1(\theta)|^2$.
$i_\parallel(\theta)$ — $|S_2(\theta)|^2$.
j_n — spherical Bessel function of order n.
J_n — cylindrical Bessel function of order n.
k — wavenumber.
$\mathbf{k}_i$ — incident wave vector.
$\mathbf{k}_s$ — scattered wave vector.
κ — index of absorption.
κ — $= 2xn_i$.
κ_{ADA} — $= 2xn_i$.
κ_{EA} — $= 2xn_r n_i$.
K_{abs} — absorption coefficient.
K_{ext} — extinction coefficient.
K_{sca} — scattering coefficient.
$KZAA$ — Kokhanovsky and Zege approximation.
λ — wavelength of the radiation.
m — $n_r + in_i$.

μ magnetic permeability/$\langle\cos\theta\rangle$.
μ_{abs} absorption coefficient.
μ_{ext} extinction coefficient.
μ_{sca} scattering coefficient.
$\mathbf{M}$ magnetic polarization.
$MADA$ modified ADA.
$MGEA$ modified GEA.
MPA main form of the Perelman approximation.
$MRGA$ modified RGA.
$MSPA$ modified SPA.
$\mathbf{N}$ refractive index of the host medium.
n_r real part of the refractive index.
n_i imaginary part of the refractive index.
NWA Nussenzweig and Wiscombe approximation.
$p(\theta)$ single scattering phase function.
ϕ single particle phase function.
ψ_n Riccati–Bessel function of order n.
$\mathbf{P}$ electric polarization.
PA Perelman approximation.
$PAPP$ Penndorf approximation.
$PSPA$ Penndorf–Shifrin–Punina approximation.
Q_{abs} absorption efficiency factor.
Q_{ext} extinction efficiency factor.
Q_{sca} scattering efficiency factor.
QSA quasi-stationary approximation.
$\tilde{\rho}_{ADA}$ $= \rho_{ADA} + i\kappa_{ADA}$.
$\tilde{\rho}_{EA}$ $= \rho_{EA} + i\kappa_{EA}$.
ρ_{ADA} $2ka(n_r - 1)$.
ρ_{EA} $= ka(n_r^2 - n_i^2 - 1)$.
RA Rayleigh approximation.
RAA Ramsauer approximation.
RGA Rayleigh–Gans approximation.
RPF Rayleigh phase function.
S_1, S_2 scattering functions for a sphere.
σ conductivity.
$\sigma(\theta, \varphi)$ differential cross section.
$SIXPPF$ six-parameter phase function.
SPA Scalar Perelman approximation.
$SPPF$ Single particle phase function.
STA Shifrin–Tonna approximation.
SRA simplified Ramsauer approximation.
θ scattering angle.
T_1, T_2 scattering functions for an infinite cylinder.
$TEWS$ transverse electric wave scattering.
$TMWS$ transverse magnetic wave scattering.
$TTHG$ two-term Henyey and Greenstein.
$TTPF$ two-term phase function.
WAA Walstra approximation.
WA Wiscombe approximation.
$WKBA$ Wentzel–Kramers–Brillouin approximation
ω angular frequency.
ω_0 albedo.
ϖ_0 co-albedo.
ξ_n Riccati–Hankel function of order n.
y_n Neumann function of order n.

Chapter 1

Introduction

1.1 Objective and scope of the book

The scattering of waves and particles by an obstacle is perhaps the most important experimental capability available to researchers for characterizing a given target. By analyzing the scattered radiation, a great deal of information about the target can be retrieved. The technique has found applications in diverse fields of studies spanning science, engineering, agriculture, medicine, atmospheric science, ocean optics, remote sensing, routine measurements in industry, and more.

Lord Rutherford (1911) investigated the structure of the atom by bombarding a thin gold foil with a beam of α particles. More than a century later, the basic idea behind current experimental resources for studying fundamental particles continues to be the same. The higher the energy of the probe, the deeper it explores the target. This is the basic logic behind scattering experiments being conducted at higher and higher energies. It is now known that the probing particles can also be looked upon as waves known as "matter waves".

Besides matter waves, there are two other wave types that are of great interest as probes. These are electromagnetic waves and acoustic waves. In a series of papers commencing 1871, Lord Rayleigh (1871) showed that the blue colour of the sky is a consequence of the scattering of sunlight by gas molecules and other small atmospheric particles. Simultaneously, he also studied the scattering of sound (Rayleigh 1872).

While the wave equations guiding propagation of matter waves, electromagnetic waves and acoustic waves have some formal similarities, they also differ from each other in a number of ways. For example, whereas electromagnetic waves are transverse vector waves, acoustic waves are longitudinal scalar waves. Matter waves are scalar waves for a spinless particle but vector wave for a spin 1 particle. All three wave types are expected to interact with matter differently. The differences occur even within the realm of a particular type of wave depending on its frequency. For example, X-rays as well as light are electromagnetic radiation. But the interaction of X-rays with a target could be at variance with how light interacts with the same target. Clearly, waves in different frequency ranges are suitable for different purposes depending on the peculiarities of the target.

One specific example of electromagnetic wave scattering, where the scattering technique has been regularly employed as a tool of inquiry, is the scattering of light. The interaction of light with a single particle (or an ensemble of particles) depends strongly on particle size (or size distribution), shape, refractive index and also on the concentration of particles in an ensemble. As a consequence, at least in principle, it should be possible to deduce information about these quantities by executing a suitable measurement of the scattered radiation. The nondestructive nature of light and the prospect of real-time study are the main features that make light scattering a preferred and popular means for exploring properties of an isolated particle as well as for particles in an ensemble. The intense interest in the topic is evident from the large number of books and monographs that have been published in the past 15–20 years on the topics relating to light scattering and the vast amount of other literature in the form of research papers and reviews.

Biomedical diagnostics and therapeutic considerations have benefited immensely from light scattering studies (see, for example, Tuchin 1997, Niemz 2003; Wax and Backman 2010, Vo-Dinh 2015). Diagnostic applications include flow cytometry for cell diagnosis and differential blood count, optical coherence tomography, optical biopsy, tissue characterization by a variety of optical measurements such as diffuse reflectance and transmittance, light scattering spectroscopy, etc. In tissues, the scattering originates from the spatial heterogeneities of the refractive index as the radiation propagates through a tissue sample. The heterogeneities can be looked upon as a consequence of different refractive indices of tissue components such as the cell nucleus and other organelles. The aim of light scattering studies is to achieve methods for decoding the differences between the normal and diseased tissue structures from the scattering measurements made on the tissue. The contributions of studies on light interaction with tissues for therapeutic use are well known as well. Photo dynamic therapy and laser surgery are two prominent examples. Both require a good understanding of light propagation in a tissue.

Extensive applications of light scattering studies occur in optical characterization of particles in flows such as in fuel sprays, agricultural sprays, domestic sprays, spray drying, spray cooling, etc. (see, for example, Xu 2000, Berrocal 2006). For injecting the fuel into a piston and gas turbine engines or coal slurry in furnaces, an appropriate spray device is required. This should produce droplets of the desired dimensions and kinematics. The objective is to maximize the energy efficiency and to minimize the emissions and residues. Medical sprays are another example where control of the drop size is crucial. The droplets to be inhaled must be in a specified size range. While larger droplets may get stuck on the way, droplets of smaller size may get expelled even before they reach their destination.

Optical methods have been used in food science and technology to determine the post harvest shelf life of fruits and vegetables. It is important that the produce be harvested at an ideal time. Light scattering methods constitute a perfect way of assessing the state of maturity, quality and ripeness of the produce in terms of its texture, colour and turbidity (see, for example, Cubeddu et al. 2002, Zude 2009, Nikolai et al. 2014, Fang et al. 2016, Lu 2016).

Material specific peculiarities often appear in nanometer-size particles. Consequently, studying the properties and interactions of nanoparticles and their aggregates has become a very active field of research in recent years. Among the numerous properties of nanoparticles, the optical properties play a vital role in determining the type of application in which a nanoparticle may be utilized or the type of nanoparticles which may be used for a desired outcome.

Besides scattering of electromagnetic waves at optical frequencies, there is interest in scattering of radio waves, microwaves, infrared, ultraviolet and X-ray regions. The wavelength range is from hundreds of meters for radio waves to 0.01 nm for X-rays. The propagation of radio waves and microwaves for the purpose of communication and remote sensing has been studied extensively for a long time. Radar and lidar-based techniques require a good understanding of radio and light wave scattering.

Much farther away from the earth's atmosphere, interstellar and interplanetary media consist of gas and dust. The information about the nature of the interstellar and interplanetary dust is critical for multiple reasons, including developing theories on the birth and death of stars. The only way to retrieve this information is by deciphering it from the electromagnetic radiation emanating from a star and reaching the observer after interaction with the interstellar medium (see, for example, Whittet 2003, Krugel 2007, Draine 2011).

A change in particulate-size distribution, say, in the atmosphere, can influence weather and climate changes. To be able to predict these changes, it is necessary to have an understanding of the interaction of electromagnetic waves with particles in the atmosphere (see, for example, Kokhanovsky 2006, 2008) and in natural water (see, for example, Jerlov 1968, Jonasz and Fournier 2007).

Many more examples can be cited. However, the array of examples provided above are sufficient to demonstrate that the problems that could possibly be addressed by way of interpreting electromagnetic wave scattering data, are innumerable and not discipline specific. Many books and monographs exist which give insight into various aspects of electromagnetic wave interaction with particles. This monograph is an attempt to focus attention on a particular aspect of the electromagnetic wave scattering problem, namely, the analytic solutions. The scope and contours of the monograph are defined below.

Broadly speaking, the solution to an electromagnetic wave scattering problem may be catalogued as analytic or numerical. The present work confines itself to analytic solutions. The analytic solutions may be further subdivided as exact and approximate solutions. While both types of solutions are important, it is the approximate solutions which generally provide direct and simple relationships between scatterer properties and the measured quantities. Hence, approximate solutions are of greater interest. New analytic solutions, exact as well as approximate, are appearing all the time in newer and newer situations, contributing to better understanding, new knowledge and increased intuition. Even the software "Mathematica" has been used to obtain analytic expressions (Matciak 2012). As demonstrated above, the scattering problems of interest occur across diverse disciplines. Identifying these and bringing together such spread out material in one place appears desirable. We see this as the primary purpose of this monograph.

The scattering problems considered in this monograph are restricted to elastic scattering processes. Thus, the scatterings such as Brillouin, Raman, Compton and processes like fluorescence lie outside the scope of this monograph. No restriction has been placed on whether the target is a single isolated scatterer or is an ensemble of particles. However, in the latter case, the constituents are always taken to be randomly separated independent scatterers.

From the target perspective, the investigations may be divided in three categories. (i) Scattering by a single particle. This approach is appropriate whenever it is possible to achieve a "one particle at a time" type of scattering arrangement; for example, in flow cytometry. (ii) Scattering by a tenuous system of uncorrelated scatterers. In such systems, multiple scattering can be ignored. The solution for such systems can be constructed by integrating a single particle scattering solution over a judicious choice of particle-size distribution—guided by the requirements of the problem. (iii) Scattering by particles in an ensemble in which multiple scatterings cannot be ignored. The problem then demands the solution of radiative transfer equation. In the present monograph, our primary focus is on scattering problems of types (i) and (ii). Only some peripheral aspects of type (iii) problems have been made part of this work.

1.2 Electromagnetic wave propagation in homogeneous media

As a first step, it is appropriate to consider the propagation of electromagnetic waves in a medium which is devoid of scatterers. This establishes the notation and vocabulary and recaps some ideas and notions which might be needed later in the monograph.

1.2.1 Dielectric medium

The macroscopic Maxwell equations in a dielectric medium with no free charges and currents are:

$$\nabla.\mathbf{D} = 0, \tag{1.1}$$

$$\nabla.\mathbf{B} = 0, \tag{1.2}$$

$$\nabla \times \mathbf{E} = -\frac{\partial \mathbf{B}}{\partial t}, \tag{1.3}$$

$$\nabla \times \mathbf{H} = \frac{\partial \mathbf{D}}{\partial t}. \tag{1.4}$$

SI units are used. The electric field vector is represented by the symbol **E**, the magnetic induction vector by **B**, the magnetic field vector by **H**, and the electric displacement vector by **D**. The linear constitutive relations are:

$$\mathbf{D} = \varepsilon_0 \mathbf{E} + \mathbf{P} = \epsilon \mathbf{E}, \tag{1.5}$$

and

$$\mathbf{B} = \mu_0\mathbf{H} + \mathbf{M} = \mu\mathbf{H}, \tag{1.6}$$

where ϵ_0 and μ_0, respectively, are the electric permittivity and the magnetic permeability of the vacuum. The corresponding quantities for the medium are ε and μ. The electric polarization $\mathbf{P}$ and the magnetization $\mathbf{M}$ are defined by the equations,

$$\mathbf{P} = \epsilon_0\chi_e\mathbf{E}\ ; \qquad \mathbf{M} = \mu_0\chi_m\mathbf{H}, \tag{1.7}$$

where χ_e and χ_m are the electric and magnetic susceptibilities, respectively. The quantity $\mathbf{P}$ gives polarization per unit volume and $\mathbf{M}$ is the magnetization per unit volume. Thus, $\varepsilon = \varepsilon_0(1+\chi_e)$ and $\mu = \mu_0(1+\chi_m)$.

The coupled Maxwell equations (1.3) and (1.4) can be decoupled by taking the curl on both sides and using the vector identity,

$$\nabla \times (\nabla \times \mathbf{A}) = \nabla(\nabla.\mathbf{A}) - \nabla.(\nabla\mathbf{A}), \tag{1.8}$$

yielding separate equations for $\mathbf{E}$ and $\mathbf{H}$:

$$\nabla^2\mathbf{E} = \mu\varepsilon\frac{\partial^2\mathbf{E}}{\partial t^2}, \tag{1.9}$$

and

$$\nabla^2\mathbf{H} = \mu\epsilon\frac{\partial^2\mathbf{H}}{\partial t^2}. \tag{1.10}$$

In the Cartesian coordinate system, each component of the $\mathbf{E}$ and $\mathbf{H}$ vectors satisfies an equation of the type,

$$\nabla^2 U = \frac{1}{v^2}\frac{\partial^2 U}{\partial t^2}, \tag{1.11}$$

which describes a wave travelling with speed v. A comparison of (1.11) with (1.9) shows that the electromagnetic wave travels in the medium with a speed,

$$v = 1/\sqrt{\varepsilon\mu} = \frac{c}{N}, \tag{1.12}$$

where N is the real refractive index of the medium given by,

$$N = \sqrt{\frac{\varepsilon\mu}{\varepsilon_0\mu_0}}. \tag{1.13}$$

For most dielectrics $\mu = \mu_0$. For such media,

$$N = \sqrt{\varepsilon/\varepsilon_0}. \tag{1.14}$$

One may also define a quantity,

$$\eta = \frac{\mu}{\varepsilon}, \tag{1.15}$$

which is called the wave impedance of the medium. For a vacuum, its value is 377 ohms. In empty space

$$v = 1/\sqrt{\varepsilon_0 \mu_0} = 3.0 \times 10^8 \text{ m/sec}, \tag{1.16}$$

is nothing but the speed of light usually denoted by c.

The solutions of (1.9) and (1.10) that are of the most practical interest are the time harmonics. That is, the solution may be written in the form,

$$\mathbf{E}(\mathbf{r}, t) = \mathbf{E}(\mathbf{r})e^{-i\omega t} \qquad \mathbf{H}(\mathbf{r}, t) = \mathbf{H}(\mathbf{r})e^{-i\omega t}, \tag{1.17}$$

where ω is the angular frequency of the monochromatic wave. Substitution from (1.17) in equations (1.9) and (1.10) then leads to the Helmholtz equations,

$$\nabla^2 \mathbf{E}(\mathbf{r}) + m^2 k^2 \mathbf{E}(\mathbf{r}) = 0, \tag{1.18}$$

for the electric field vector and

$$\nabla^2 \mathbf{H}(\mathbf{r}) + m^2 k^2 \mathbf{H}(\mathbf{r}) = 0, \tag{1.19}$$

for the magnetic field vector, with $k^2 = \mu\epsilon\omega^2 = \omega^2/c^2$, k being the wave number. The plane wave solutions of (1.18) and (1.19) are well known. Their substitution in (1.17) leads to:

$$\mathbf{E}(\mathbf{r}, t) = \mathbf{E}_0 e^{i(\mathbf{k}.\mathbf{r} - \omega t)}, \tag{1.20}$$

and

$$\mathbf{H}(\mathbf{r}, t) = \mathbf{H}_0 e^{i(\mathbf{k}.\mathbf{r} - \omega t)}, \tag{1.21}$$

where $\mathbf{E}_0$ and $\mathbf{H}_0$ are (complex) amplitudes. The wave vector $\mathbf{k}$ expresses the propagation direction. It is tacitly assumed that the actual solution of the wave equation is given by the real components on the right-hand sides of (1.20) and (1.21). The complex number representation is used because it makes the handling of mathematical expressions and the accompanying algebra simpler. At any given time t, surfaces $\mathbf{k}.\mathbf{r}$ = constant are planes of constant phase and hence the wave is called a plane wave.

An alternative convention is often used, in which a time harmonic plane wave is represented as $\exp(-i(\mathbf{k}.\mathbf{r} - \omega t))$. Note that i in (1.20) has been replaced by $(-i)$. This convention for representing a plane wave appears in the books by van de Hulst (1957), Kerker (1969) and Kokhanovsky (2004). The convention adopted here is that used in the books by Newton (1966), Bayvel and Jones (1981), Bohren and Huffman (1983), Mishchenko et al. (2002), Babenko et al. (2003), and Sharma and Somerford (2006). In this notation the complex refractive index is expressed as $m = n_r + in_i$. The scattering function is the complex conjugate of that in the other notation. The detailed implications of selecting a particular convention were discussed in Shifrin and Zolotov (1993).

With the solutions given by (1.20) and (1.21), the Maxwell equations (1.1) to (1.4) impose the following added restrictions on the fields,

$$\nabla.\mathbf{E} = 0 \quad \rightarrow \quad \mathbf{k}.\mathbf{E} = 0, \tag{1.22}$$

$$\nabla.\mathbf{B} = 0 \quad \rightarrow \quad \mathbf{k}.\mathbf{B} = 0, \tag{1.23}$$

$$\nabla \times \mathbf{E} = -\frac{\partial \mathbf{B}}{\partial t} \quad \rightarrow \quad \mathbf{k} \times \mathbf{E} = \omega \mathbf{B}, \tag{1.24}$$

$$\nabla \times \mathbf{H} = \frac{\partial \mathbf{D}}{\partial t} \quad \rightarrow \quad \mathbf{k} \times \mathbf{H} = -\varepsilon\omega \mathbf{E}, \tag{1.25}$$

implying that **E** and **H** are perpendicular to the direction of propagation vector **k** and to each other.

Further, by writing

$$\mathbf{E_0} = \hat{\mathbf{n}} E_0 e^{i\delta}, \tag{1.26}$$

the electric and magnetic field vectors may be expressed as,

$$\mathbf{E}(\mathbf{r}, t) = E_0 e^{i(\mathbf{k}.\mathbf{r} - \omega t + \delta)} \hat{\mathbf{n}}, \tag{1.27}$$

and

$$\mathbf{H}(\mathbf{r}, t) = \frac{k}{\omega \mu} (\hat{\mathbf{k}} \times \hat{\mathbf{n}}) E_0 e^{i(\mathbf{k}.\mathbf{r} - \omega t + \delta)}, \tag{1.28}$$

respectively. The unit vector $\hat{\mathbf{n}}$ determines the direction of the electric field vector, E_0 is a real number and $e^{i\delta}$ is a complex phase. It may be noted that **E** and **H** are in phase.

The flux density transported by the wave is given by the Poynting vector,

$$\mathbf{S} = (\mathbf{E} \times \mathbf{H}). \tag{1.29}$$

At the frequencies involved ($\sim 10^{15}$ Hz in the visible region), the variation of **S** with time is very rapid. Most instruments are not capable of responding to such fast changes. Therefore, what one actually measures is the average power per unit area, I, called intensity. The temporal averaging is over a time T, such that $\omega T \gg 1$. Under this condition, the intensity can be readily shown to be:

$$I = |\langle \mathbf{S} \rangle| = \frac{1}{2} \varepsilon v E_0^2, \tag{1.30}$$

where the contributions of the electric and magnetic fields are equal.

If $\varepsilon(\mathbf{r}) \neq$ constant spatially, then $\nabla.\mathbf{E} \neq 0$. Instead, $\nabla.(\varepsilon(\mathbf{r})\mathbf{E}) = 0$. This yields,

$$\nabla.\mathbf{E} = -\frac{1}{\varepsilon(\mathbf{r})} \mathbf{E}.\nabla\varepsilon(\mathbf{r}). \tag{1.31}$$

The wave equation for the electric field then becomes,

$$\nabla^2\mathbf{E}(\mathbf{r}) + \nabla\left(\mathbf{E}(\mathbf{r}).\nabla \log\varepsilon(\mathbf{r})\right) + k^2\varepsilon(\mathbf{r})\mathbf{E}(\mathbf{r}) = 0. \tag{1.32}$$

The second term is zero only if the spatial variation of $\epsilon(\mathbf{r})$ or $m(\mathbf{r})$ is zero.

1.2.2 Conducting medium

In a conducting medium, equation (1.4) is modified to,

$$\nabla \times \mathbf{H} = \frac{\partial \mathbf{D}}{\partial t} + \sigma\mathbf{E}, \tag{1.33}$$

where σ is the conductivity of the medium. The wave equations for electric and magnetic fields in such a medium can be obtained exactly in the same manner as in the case of a non-conducting medium. The modified wave equations still allow a time harmonic plane wave solution but the propagation vector now has an imaginary part too:

$$\mathbf{k} = \mathbf{k}' + i\mathbf{k}'', \tag{1.34}$$

where

$$|\mathbf{k}'| = \omega\sqrt{\frac{\varepsilon\mu}{2}}\left[\sqrt{1+\left(\frac{\sigma}{\varepsilon\omega}\right)^2}+1\right]^{1/2}, \tag{1.35}$$

and

$$|\mathbf{k}''| = \omega\sqrt{\frac{\varepsilon\mu}{2}}\left[\sqrt{1+\left(\frac{\sigma}{\varepsilon\omega}\right)^2}-1\right]^{1/2}. \tag{1.36}$$

The refractive index and fields corresponding to the propagation vector defined in (1.35) to (1.36) are:

$$n_r = \sqrt{\frac{\varepsilon\mu}{2}}\left[\sqrt{1+\left(\frac{\sigma}{\varepsilon\omega}\right)^2}+1\right]^{1/2}, \tag{1.37}$$

$$n_i = \sqrt{\frac{\varepsilon\mu}{2}}\left[\sqrt{1+\left(\frac{\sigma}{\varepsilon\omega}\right)^2}-1\right]^{1/2}, \tag{1.38}$$

$$\mathbf{E}(\mathbf{r},t) = \mathbf{E}_0 e^{-\mathbf{k}''.\mathbf{r}} e^{i(\mathbf{k}'.\mathbf{r}-\omega t)}, \tag{1.39}$$

and

$$\mathbf{H}(\mathbf{r},t) = \mathbf{H}_0 e^{-\mathbf{k}''.\mathbf{r}} e^{i(\mathbf{k}'.\mathbf{r}-\omega t+\phi_l)}. \tag{1.40}$$

Equations (1.39) and (1.40) together represent a plane electromagnetic wave propagating in the direction given by wave vector $\mathbf{k}'$. The wave decays in amplitude in the direction given by $\mathbf{k}''$. If $\mathbf{k}'$ and $\mathbf{k}''$ are parallel, the wave is homogeneous; else, it is an inhomogeneous wave. The magnetic vector lags behind the electric vector and contributes more to the energy density in comparison to the electric field.

1.2.3 Optically active medium

A model of an optically active medium consists of uniformly distributed chiral objects. A chiral object, by definition, is any geometrical figure which cannot be superimposed on its mirror image.

The constitutive relations for such a medium depend on the specific model adopted. All researchers do not agree on a single model. A widely accepted model consists of lossless short wire helices of same handedness (see, for example, Jaggard et al. 1979, Bassiri 1987, Bohren and Huffman 1983). For time harmonic waves, constitutive relations in this model can be written as,

$$\mathbf{D} = \epsilon \mathbf{E} + \gamma\varepsilon(\nabla \times \mathbf{E}), \tag{1.41}$$

and

$$\mathbf{H} = -\beta\,(\nabla \times \mathbf{H}) + \frac{1}{\mu}\mathbf{B}, \tag{1.42}$$

where γ and β are the cross susceptibilities. Further, if it is assumed that $\gamma = \beta \equiv \gamma$, (1.41) and (1.42) become,

$$\mathbf{D} = \varepsilon \mathbf{E} + \gamma\varepsilon\,(\nabla \times \mathbf{E})\,, \tag{1.43}$$

and

$$\mathbf{H} = -\gamma\,(\nabla \times \mathbf{H}) + \frac{1}{\mu}\mathbf{B}. \tag{1.44}$$

For $\gamma > 0$ the medium is right handed, for $\gamma < 0$ it is left handed and for $\gamma = 0$ it is achiral. Other often quoted constitutive relations are from Condon (1937) and Born (1972) and can be found in a review of optical properties of chiral media by Lenekar (1996).

With constitutive relations given by (1.43) and (1.44), the Maxwell equation

$$\nabla \times \mathbf{H} = -i\omega \mathbf{D}, \tag{1.45}$$

becomes

$$\nabla \times \left(-\gamma(\nabla \times \mathbf{H}) + \frac{1}{\mu}\mathbf{B}\right) = i\omega\,(\varepsilon \mathbf{E} + \gamma\varepsilon\nabla \times \mathbf{E})\,, \tag{1.46}$$

which can also be expressed as,

$$(1 - \omega^2\gamma^2\varepsilon\mu)\nabla^2\mathbf{E} + k^2\mathbf{E} + 2\omega\varepsilon\mu\gamma\nabla \times \mathbf{E} = 0. \tag{1.47}$$

Substitution of a plane wave solution of the form

$$\mathbf{E}(\mathbf{r}) = \left(\hat{\mathbf{e}}_\perp E_\perp + \hat{\mathbf{e}}_\parallel E_\parallel\right)e^{ihz}, \tag{1.48}$$

in (1.47) leads to simultaneous equations for $\mathbf{E}_\perp$ and $\mathbf{E}_\parallel$ which can be solved to yield two physically permissible solutions:

$$h_1 \equiv k_L = \omega\sqrt{\mu\varepsilon}\left[\frac{1 + \gamma\omega\sqrt{\varepsilon\mu}}{1 - \gamma^2\omega^2\varepsilon\mu}\right], \tag{1.49}$$

and

$$h_2 \equiv k_R = \omega\sqrt{\mu\varepsilon}\left[\frac{1-\gamma\omega\sqrt{\varepsilon\mu}}{1-\gamma^2\omega^2\varepsilon\mu}\right]. \tag{1.50}$$

The orthogonal unit vectors $\hat{\mathbf{e}}_{perp}$ and $\hat{\mathbf{e}}_{parallel}$ are perpendicular and parallel, respectively, to the scattering plane. The direction of propagation has been taken to be along the $z-$axis and h is the propagation constant. The solution (1.49) corresponds to a wave travelling with phase velocity ω/k_L and (1.50) corresponds to a wave travelling with a phase velocity ω/k_R. Subscripts L and R correspond to the left and right circularly polarized waves. It can be checked that

$$\gamma = \frac{1}{2}\left(\frac{1}{k_R} - \frac{1}{k_L}\right). \tag{1.51}$$

For $k_R = k_L$, $\gamma = 0$. That is, the medium becomes achiral.

The two components of a linearly polarized wave, the left circularly polarized wave and the right circularly polarized wave, have different phase velocities inside the chiral medium. As a result, the plane of polarization of the transmitted wave gets rotated with respect to the plane of polarization of the incident linearly polarized wave. The amount of rotation is proportional to the distance travelled by the wave in the slab. If the two circular components are absorbed to different extents, the result is known as the circular dichroism.

1.2.4 Anisotropic medium

In an isotropic medium, the dielectric permittivity is a scalar quantity. Consequently, the displacement vector $\mathbf{D}$ and the field vector $\mathbf{E}$ are in the same direction. In contrast, in an anisotropic medium, $\mathbf{D}$ and $\mathbf{E}$ need not be parallel. Thus, an electric field applied in the x direction results in a $\mathbf{D}$ which has all three components:

$$D_x = \varepsilon_{xx}E_x, \quad D_y = \varepsilon_{yx}E_x, \quad D_z = \varepsilon_{zx}E_x. \tag{1.52}$$

Thus, the general form of the constitutive relation can be expressed in the following matrix form:

$$\begin{pmatrix} D_x \\ D_y \\ D_z \end{pmatrix} = \begin{pmatrix} \varepsilon_{xx} & \varepsilon_{xy} & \varepsilon_{xz} \\ \varepsilon_{yx} & \varepsilon_{yy} & \varepsilon_{yz} \\ \varepsilon_{zx} & \varepsilon_{zy} & \varepsilon_{zz} \end{pmatrix} \begin{pmatrix} E_x \\ E_y \\ E_z \end{pmatrix}. \tag{1.53}$$

It has been shown that a coordinate system, known as a principal axis system, can always be chosen in which it is possible to draft the dielectric tensor in a diagonal form:

$$\begin{pmatrix} D_x \\ D_y \\ D_z \end{pmatrix} = \begin{pmatrix} \varepsilon_{xx} & 0 & 0 \\ 0 & \varepsilon_{yy} & 0 \\ 0 & 0 & \varepsilon_{zz} \end{pmatrix} \begin{pmatrix} E_x \\ E_y \\ E_z \end{pmatrix}. \tag{1.54}$$

If $\varepsilon_{xx} \neq \varepsilon_{yy} \neq \varepsilon_{zz}$, the medium is biaxial. If $\varepsilon_{xx} = \varepsilon_{yy} \neq \varepsilon_{zz}$, the medium is uniaxial. A consequence of the fact that **D** and **E** are not aligned is that while **D** is perpendicular to the propagation vector, **S** and **E** are not perpendicular to **k** except when **E** is along one of the principal axes. Also, because the velocity of components along different axes are different, the phase of the wave changes continuously. For example, a birefringent plate modifies the phase of incident wave by

$$\delta = \pm \frac{2\pi(n_e - n_o)d}{\lambda}, \tag{1.55}$$

where d is the thickness of the plate and n_e and n_0 are the refractive indices seen by two orthogonally polarized waves. If **D** depends on **E** as well as on **H**, the medium is called bianisotropic.

1.3 Classification of electromagnetic scattering problems

If an obstacle (single scatterer or a cloud of particles) is introduced in the path of the wave, the wave particle interaction depends on the type of obstacle and the energy of the wave. A possible classification of electromagnetic scattering problems has been sketched as a detailed flowchart in the book by Mishchenko et al. (2006). For our purpose, we define below some scattering types which require mention in this work.

1.3.1 Wave, particle and ray descriptions

In some problems it is easier to express ideas in terms of photon language, while in some other situations it is found to be more convenient to express notions in the language of waves. All quanta are packets of energy. They differ from one another in their energy and momentum. The energy E, and the momentum p, of a photon are related to the frequency ν of the corresponding wave as,

$$E = h\nu, \tag{1.56}$$

and

$$p = E/c = h/\lambda, \tag{1.57}$$

where h is the well-known Planck's constant and λ is the wavelength of the corresponding wave.

In a ray picture, virtual rays are imagined which are paths along which the energy is transported. Simple laws governing reflection and refraction of electromagnetic waves determine the trajectory of rays. This picture is particularly useful in describing the interaction of electromagnetic waves with large scatterers. That is, when $ka = 2\pi a/\lambda \gg 1$, with a being the characteristic size of the scatterer.

1.3.2 Elastic, quasi-elastic and inelastic scattering

If the frequency of a scattered wave is the same as that of an incident wave the scattering is called elastic. The wave may change its direction and polarization; but no energy is lost. Rayleigh and Mie scatterings are examples of elastic scattering. On the other hand, if the frequency of the outgoing wave is different from that of the incident wave, the scattering is inelastic. Raman, Brillouin and Compton scatterings are some examples of inelastic scattering.

Quasi-elastic scattering involves a minute frequency shift in the frequency of the incident electromagnetic wave. This shift is a consequence of the interaction of incident radiation with moving scatterers (Doppler shift).

1.3.3 Static and dynamic scattering

In static light scattering (SLS), intensity averaged over a period of time that is greater than $1/\omega$ is measured. In contrast, dynamic light scattering (DLS) measures real-time intensities. The DLS measurements are particularly useful for characterization of particles in an ensemble. At any given instant of time, the scatterers in an ensemble may be treated as static. The scattered fields from several randomly placed scatterers interfere constructively or destructively. A pattern of bright and dark spots results in the far field. This is the speckle pattern. A change in particle positions leads to change in the speckle pattern which can be analyzed to yield information about particles in an ensemble. Phase correlation spectroscopy (PCS) is an alternative name for the DLS.

1.3.4 Single and multiple scattering

The term single particle scattering refers to scattering by an isolated particle. On the other hand, the term single scattering (distinct from single particle scattering) is assigned exclusively for use in the context of scattering by a particle in an ensemble. If a photon in the incident beam encounters only one scattering before arriving at the detector, the scattering is labelled as single scattering. If the incident photon encounters more than one particle on its way to the detector, the scattering is multiple scattering. The extent of multiple scattering depends on the turbidity of the medium; that is, on the concentration of scatterers. Various estimates of conditions for neglect of multiple scattering exist in the literature. Kerker (1969) gives the criterion that the turbidity should be less than 0.1 or the transmission of photons from the collection of particles is more than about 90%. Alternative considerations suggest this transmission should be more than 60% (see, for example, Bayvel and Jones 1981). A comparatively recent study by Mishchenko et al. (2007) for wavelength-sized particles concludes that the packing fraction of scatterers should be less than 1% and $kd < 30$ for the multiple scattering to be neglected. Here, d is the average separation between scatterers.

1.3.5 Independent and dependent scattering

If the scattering characteristics of a particle in a cloud of particles are independent of separation with other particles, the scattering is independent. In other words, the presence of surrounding particles does not influence the scattering characteristics of the scattering particle. It scatters the wave in a way that it would have done if other particles were absent.

As the particles come closer, their electric fields interact directly. Estimates of the separation between particles exceeding that point which the particles may be treated as independent scatterers differ. While Kerker (1969) gives this separation to be equal to about 3 diameters, Mishchenko et al. (1996) estimate this for wavelength-sized spheres to be about four diameters. In terms of volume fraction, 3% seems to be the upper limit for scattering to be independent (see, for example Jones 1999).

The dependent scattering problems are not included here as a specific theme. The topic has been discussed by many authors. Some of the works useful for further consultation include Ivanov (1988), Gobel et al. (1995), Kohanovsky (2006), and Mishchenko et al. (2006), among many others.

1.3.6 Rayleigh scattering

The term Rayleigh scattering is associated with scattering by particles much smaller in size than the wavelength of the incident radiation. That is, $ka \ll 1$. As the size of the scatterer is small compared to the wavelength, the electric field of the wave generates an oscillating dipole by separating the electron cloud and the nucleus. The oscillating dipole then re-radiates. The frequency of the radiation is obviously the same as the frequency of the incident wave. The cross section for this elastic scattering process is proportional to $1/\lambda^4$. Rayleigh explained the blue colour of the sky on the basis of the scattering of sunlight by such small scatterers and hence the name Rayleigh scattering.

1.3.7 Mie scattering

The term Mie scattering is used when the size of the scattering particle is on the order of a wavelength or larger. Strictly speaking, however, Mie scattering refers to the exact solution of the problem of electromagnetic wave scattering by a homogeneous sphere (Mie 1908). The cross section for Mie scattering is proportional to $1/\lambda^2$ for a sphere. Mie scattering too belongs to the elastic scattering class.

1.4 Single particle scalar scattering

1.4.1 Basic definitions

Consider a beam of a scalar wave in a nonabsorbing medium incident on an obstacle. The characteristic size and the complex relative refractive index, with respect to the host medium, are a and $m = n_r + in_i$, respectively. For a sphere, a is its radius. Let the wavelength of the incident wave be λ. The propagation vectors of incident and scattered waves are denoted by $\mathbf{k}_i$ and $\mathbf{k}_s$, respectively. The scattering under consideration is elastic. Therefore, $|\mathbf{k}_i| = |\mathbf{k}_s| = k$. A schematic scattering diagram is shown in Figure 1.1. The detector is at a distance r from the scatterer. The wave equation for a scalar field $U(\mathbf{r})$ may then be written as,

$$\nabla^2 U(\mathbf{r}) + k^2 m^2(\mathbf{r}) U(\mathbf{r}) = 0, \tag{1.58}$$

where $m(\mathbf{r}) = m$ inside the scatterer and $m(\mathbf{r}) = 1$ outside the scatterer.

The total field in the far field region ($r \gg a$) is a sum of the incident plane wave field,

$$U_{inc} = e^{ikz - i\omega t}, \tag{1.59}$$

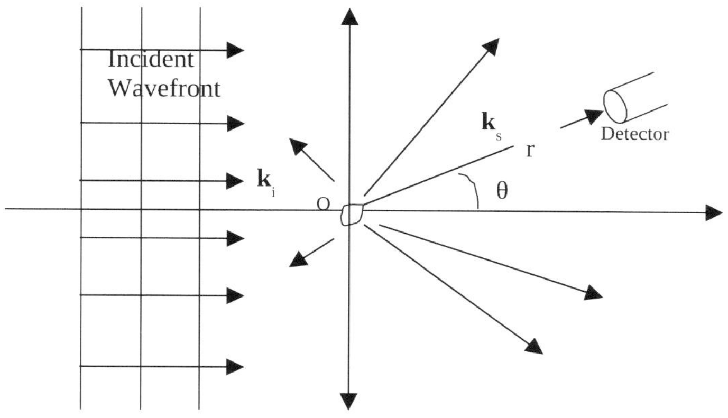

FIGURE 1.1: A schematic diagram showing scattering of a scalar wave.

and the added scattered field,

$$U_{sca} = \frac{e^{ikr-i\omega t}}{ikr} S(\theta, \varphi), \tag{1.60}$$

which is a spherical wave. An extra factor k has been introduced here in the denominator to make the scattering amplitude $S(\theta, \varphi)$ a dimensionless quantity. The complex amplitude $S(\theta, \varphi)$ is sometimes also referred to as the scattering function . The subscripts *inc* (or *i*) and *sca* (or *s*), respectively, denote incident and scattered quantities throughout this monograph. The scattering angle is θ, and φ is the azimuthal angle. Equation (1.60) may be expressed in terms of the incident field as,

$$U_{sca} = \frac{e^{ikr-ikz}}{ikr} S(\theta, \varphi) U_{inc} \, . \tag{1.61}$$

Intensity being proportional to the square modulus of the amplitude, (1.61) yields,

$$\frac{I_{sca}}{I_{inc}} = \frac{|S(\theta, \varphi)|^2}{k^2 r^2} = \frac{i(\theta, \varphi)}{k^2 r^2}, \tag{1.62}$$

where

$$i(\theta, \varphi) \equiv |S(\theta, \varphi)|^2 \; = \; k^2 \sigma_{sca}(\theta, \varphi), \tag{1.63}$$

with $\sigma_{sca}(\theta, \varphi)$ as the differential scattering cross section. $i(\theta, \varphi)$ is the scattered irradiance per unit incident irradiance. This has been referred to as the scattered intensity in this book. The integration of differential cross section over all directions gives

$$C_{sca} = \int_0^{2\pi} \int_0^{\pi} \frac{i(\theta, \varphi)}{k^2} \sin\theta d\theta d\varphi, \tag{1.64}$$

which is the total scattering cross section . It represents the power removed from the incident beam in terms of beam area by the process of scattering ($P_{sca} = C_{sca} I_{inc}$). The irradiance I is the energy crossing per unit time, per unit area.

In addition to the power removed by scattering, some energy is absorbed by the material and is later emitted as heat energy at lower wavelengths. The cross section for the absorption process is denoted by C_{abs}. The total energy removed from the beam is called the extinction cross section C_{ext},

$$C_{ext} = C_{abs} + C_{sca}. \tag{1.65}$$

One also defines efficiency factors for scattering, absorption and extinction as,

$$Q_{sca} = \frac{C_{sca}}{G}, \quad Q_{abs} = \frac{C_{abs}}{G}, \quad Q_{ext} = \frac{C_{ext}}{G}, \tag{1.66}$$

where G is the geometrical cross section of the particle. The extinction efficiency is related to the real part of the scattering function via the relation,

$$Q_{ext} = \frac{4}{x^2}\mathrm{Re}[S(0)], \tag{1.67}$$

which is known as "extinction theorem" in optics and electromagnetic scattering. Analogous relations in nuclear scattering and quantum mechanics are cited as "optical theorem". The theorem is valid only if the scattering function satisfies the unitarity property.

Another specification of interest in scattering problems is the single scattering albedo. It is defined as,

$$\omega_0 = \frac{Q_{sca}}{Q_{ext}} = 1 - \frac{Q_{abs}}{Q_{ext}}. \tag{1.68}$$

For a highly scattering object $Q_{sca} \gg Q_{abs}$, and hence $\omega_0 \to 1$. On the other hand, for a large totally absorbing sphere $Q_{ext} = 2Q_{abs}$. Hence, $\omega_0 = 1/2$. A related quantity is co-albedo. It is specified as $\varpi_0 = 1 - \omega_0 = Q_{abs}/Q_{ext}$. This parameter is 0 when $Q_{abs} = 0$ and 1/2 for a totally absorbing sphere.

Sometimes it is more advantageous to depict the differential scattering cross section in terms of a parallel quantity known as the scattering phase function. It is defined as,

$$\phi(\theta,\varphi) = \frac{|S(\theta,\varphi)|^2}{k^2 C_{sca}} = \frac{\sigma_{sca}}{C_{sca}}, \tag{1.69}$$

and obeys the following normalization condition

$$\int_0^{2\pi}\int_0^{\pi} \phi(m,k,a,\theta,\varphi)\sin\theta d\theta d\varphi = 1. \tag{1.70}$$

This definition of phase function allows it to be interpreted as giving the probability of a photon being scattered in the direction (θ,φ). It is relevant to clarify that the word "phase" has no relation with the phase of the electromagnetic wave. It originates from the astronomy literature where it refers to lunar phases (Ishimaru 1999).

A plot of phase function, in general, is not symmetric about $\theta = 90$ deg. The measure of this asymmetry is denoted by the parameter g. Mathematically, the asymmetry parameter is expressed as,

$$g = \frac{\int_0^{\pi} \phi(m,k,a,\theta,\varphi)\sin\theta\cos\theta d\theta}{\int_0^{\pi} \phi(m,k,a,\theta,\varphi)\sin\theta d\theta}. \tag{1.71}$$

The values $g = 1$, $g = 0$ and $g = -1$ correspond to forward-peaked, symmetric and backward-peaked scatterings, respectively. The asymmetry parameter rapidly increases from close to zero to near 1 as ka increases and then becomes nearly constant.

1.4.2 Scalar wave scattering versus potential scattering

The nonrelativistic Schroedinger equation for a particle of mass M in the presence of a time independent potential $V(\mathbf{r})$ of range a may be written as,

$$\left[\frac{-\hbar^2}{2M}\nabla^2 + V(\mathbf{r})\right]\psi(\mathbf{r}) = E\psi(\mathbf{r}), \tag{1.72}$$

where the particle energy E is $\hbar^2k^2/2M$ and $\psi(\mathbf{r})$ is the wave function of the particle. A comparison of (1.72) with (1.58) shows that the two equations are formally similar. Thus, by identifying

$$V(\mathbf{r}) = \frac{\hbar^2k^2}{2M}\left[1 - m^2(\mathbf{r})\right], \tag{1.73}$$

the scattering picture developed for potential scattering can be used for scattering of electromagnetic waves and vice versa. Note that the equivalent potential is now energy dependent.

The analogy, however, does not extend to time dependent interactions. The time dependent Schroedinger equation that governs the interaction of matter waves with a time dependent potential is,

$$\left[\frac{-\hbar^2}{2M}\nabla^2 + V(\mathbf{r},\mathbf{t})\right]\psi(\mathbf{r},\mathbf{t}) = i\hbar\frac{\partial\psi(\mathbf{r},\mathbf{t})}{\partial t}. \tag{1.74}$$

This is a parabolic differential equation. In contrast, the scalar wave equation (1.58) and the Schroedinger equation (1.72) are hyperbolic differential equations. This difference, even at the level of wave equations, suggests that there is no reason to expect the analogy, seen for time independent Schroedinger equations, to hold for time dependent interactions.

The Schroedinger equation (1.72) can also be cast as an integral equation known as the Lippmann–Schwinger equation ,

$$\psi(\mathbf{r}) = \exp\left(i\mathbf{k}_i.\mathbf{r}\right) - \frac{2M}{\hbar^2}\int G(\mathbf{r}-\mathbf{r}')V(\mathbf{r}')\psi(\mathbf{r}')\mathbf{dr}', \tag{1.75}$$

where

$$G(\mathbf{r}-\mathbf{r}') = \frac{\exp(i\mathbf{k}|\mathbf{r}-\mathbf{r}'|)}{4\pi|\mathbf{r}-\mathbf{r}'|}, \tag{1.76}$$

is the free particle Green's function. From (1.75) and (1.76), it can be shown that the scattering amplitude $f(\theta,\varphi)$ is then given by the expression,

$$f(\theta,\varphi) = -\frac{M}{2\pi\hbar^2}\int e^{-i\mathbf{k}_s.\mathbf{r}}V(\mathbf{r})\psi(\mathbf{r})d\mathbf{r}. \tag{1.77}$$

Hence, if $\psi(\mathbf{r})$ can be determined inside the region $V(\mathbf{r}) \neq 0$, the scattering amplitude can be computed using (1.77). The original problem thus reduces to finding the wave function inside the scattering region. Using the analogy in (1.73), the scattering amplitude for a scalar electromagnetic wave may be written as,

$$S(\theta,\varphi) = -ikf(\theta,\varphi) = \frac{ik^3}{4\pi}\int e^{-i\mathbf{k}_s.\mathbf{r}}\Big[1 - m^2(\mathbf{r})\Big]\psi(\mathbf{r})d\mathbf{r}. \tag{1.78}$$

Therefore, at the level of time independent interactions, the problem of the scattering of scalar electromagnetic waves by an obstacle is analogous to the scattering of a particle of mass M by a potential. Clearly, the results obtained in one context can be translated to the other context by employing (1.73). For a sphere, $m(\mathbf{r}) = m$, when $a \leq r$, and $m = 1$ otherwise.

If the scattering potential is a central potential, it is possible to express the scattering amplitude as a partial wave sum (see, for example, Schiff 1968),

$$f(\theta) = \frac{1}{2ik}\sum_{l=0}^{l=\infty}(2l+1)\left[e^{2i\delta_l} - 1\right]P_l(\cos\theta), \tag{1.79}$$

where $P_l(\cos\theta)$ is the l-th order Legendre polynomial and δ_l is the phase shift suffered by the l-th partial wave given by the relation,

$$\frac{e^{2i\delta_l} - 1}{2} \equiv b_l = \frac{m\psi_l(x)\psi_l'(mx) - \psi_l'(x)\psi_l(mx)}{m\psi_l'(mx)\xi_l(x) - \psi_l(mx)\xi_l'(x)}, \tag{1.80}$$

where $\psi_l(x) = xj_l(x)$, and $\xi_l(x) = xh_l^{(1)}(x)$ are Riccati-Bessel and Riccati-Hankel functions with $h_l^{(1)}(x) = j_l(x) + iy_l(x)$, $y_l(x)$ being the Neumann function. Similar relations apply when x changes to mx.

In a semiclassical description of scattering, it is possible to interpret plane waves in terms of particle beam. The l-th partial wave can be associated with an impact parameter b. The relation between l and b is (see, for example, van de Hulst 1957, Nussenzweig 1992),

$$l + 1/2 = kb. \tag{1.81}$$

The impact parameter b (see Figure 3.6) is the distance in the $x-y$ plane from the centre of a beam propagating along the $z-$axis. The relationship (1.81) is valid only if $ka \gg 1$.

The problem of scattering of scalar waves by an infinitely long cylinder at perpendicular incidence is essentially a two-dimensional version of the problem of the scattering by a sphere. The scattering function can then be written as,

$$T(\theta) = \frac{ik^2}{4}\int e^{-i\mathbf{k}_s.\mathbf{b}}\left[1 - m^2(\mathbf{b})\right]\psi(\mathbf{b})d\mathbf{b}. \tag{1.82}$$

where $\mathbf{b}$ is a two-dimensional vector in the (x, y) plane, z being the direction of the cylinder axis.

1.4.3 Applicability of the scalar approximation

The electromagnetic scattering problem has often been approximated as a scalar wave scattering problem. The errors introduced by the scalar approximation were assessed by Sharma et al. (1981, 1982) by computing the ratio $i(0)_{exact}/i(0)_{scalar}$ for the scattering of:

(i) Unpolarized light by a homogeneous sphere and
(ii) Transverse electric wave scattering (TEWS) by an infinitely long homogeneous cylinder.

It was concluded that for near-forward scattered intensities, the approximation of light by a scalar wave is a reasonably good approximation if $x \geq 3.0$. For example, the ratio $i(0)_{exact}/i(0)_{scalar}$ at $x = 3$ is ≈ 0.9 for a sphere as well as for an infinitely long cylinder. For higher x, the ratio rapidly approaches unity. The relative refractive index in this computation was $m = 1.15$. For transverse magnetic wave scattering (TMWS) the electromagnetic wave equation is identical to the scalar wave equation. That is there is no error due to scalar approximation for TMWS.

A study to examine the validity of the scalar approximation for the case of a spherically symmetric continuously varying dielectric scatterer was done by Arnush (1964). On the basis of this study it was concluded that the scalar approximation is accurate if the inequality

$$\sqrt{\varepsilon(r)}\frac{d}{dr}\frac{1}{\sqrt{\varepsilon(r)}} \ll k^2\varepsilon(r), \tag{1.83}$$

is satisfied for all r.

1.5 Vector description

A scalar wave of given frequency may be specified completely by its intensity. But, a vector wave needs additional parameters for its specification. There are a number of ways to accomplish this specification. The most widely used is a set of parameters known as the Stokes parameters.

1.5.1 Stokes parameters

Let us represent the electric vector of a plane electromagnetic wave as:

$$\mathbf{E} = \left(E_{\parallel}\hat{\mathbf{e}}_{\parallel} + E_{\perp}\hat{\mathbf{e}}_{\perp}\right)e^{i(kz-\omega t)}, \tag{1.84}$$

with

$$E_{\parallel} = E_{0\parallel} e^{-i\delta_{\parallel}}, \tag{1.85}$$

and

$$E_{\perp} = E_{0\perp} e^{-i\delta_{\perp}}, \tag{1.86}$$

The $z-$axis has been chosen along the direction of propagation. Unit vectors $\hat{\mathbf{e}}_{\parallel}$ and $\hat{\mathbf{e}}_{\perp}$ are such that $\hat{\mathbf{e}}_{\perp} \times \hat{\mathbf{e}}_{\parallel} = \hat{\mathbf{z}}$.

The Stokes parameters are then defined as,

$$I = E_{\parallel} E_{\parallel}^* + E_{\perp} E_{\perp}^* = E_{0\parallel}^2 + E_{\perp}^2, \tag{1.87}$$

$$Q = E_{\parallel} E_{\parallel}^* - E_{\perp} E_{\perp}^* = E_{0\parallel}^2 - E_{\perp}^2, \tag{1.88}$$

$$U = E_{\parallel} E_{\perp}^* + E_{\perp} E_{\parallel}^* = 2E_{0\parallel} E_{0\perp} \cos\delta, \tag{1.89}$$

$$V = i(E_{\parallel} E_{\perp}^* - E_{\perp} E_{\parallel}^*) = 2E_{0\parallel} E_{0\perp} \sin\delta, \tag{1.90}$$

where $\delta = \delta_{\parallel} - \delta_{\perp}$ and the asterisk denotes the complex conjugate quantity. Physical interpretation and measurement of these parameters has been discussed in many books including Bohren and Huffman (1983) and Guenther (1990). The component I is the total flux or the intensity. The component Q is the difference between the intensity transmitted through a linear polarizer along the parallel (horizontal) and perpendicular (vertical) axes, respectively. The component U is the difference between flux transmitted through a linear polarizer at 45^0 and 135^0 to the parallel axis. Finally, V is the difference between flux transmitted through a right circular polarizer and a left circular polarizer.

1.5.2 Scattering matrix

The scalar wave scattering formalism described in Section (1.4.1) can be generalized to electromagnetic wave scattering by rewriting (1.61) as,

$$\begin{pmatrix} E_{\parallel}^s \\ E_{\perp}^s \end{pmatrix} = \frac{e^{ik(r-z)}}{ikr} \begin{pmatrix} S_2 & S_3 \\ S_4 & S_1 \end{pmatrix} \begin{pmatrix} E_{\parallel}^i \\ E_{\perp}^i \end{pmatrix}. \tag{1.91}$$

The direction of polarization of the incident and scattered waves is relative to the plane of measurement which contains incident and scattering waves. The complex elements S_j of the amplitude matrix, in general, depend on the scattering angle θ and azimuthal angle φ. If the particle is spherically symmetric, for example, a sphere, $S_3 = S_4 = 0$ and the elements S_1 and S_2 are functions of θ alone.

The relation between incident and scattered Stokes vectors follows from the amplitude scattering matrix,

$$\begin{pmatrix} I \\ Q \\ U \\ V \end{pmatrix} = \frac{1}{k^2 r^2} \begin{pmatrix} S_{11} & S_{12} & S_{13} & S_{14} \\ S_{21} & S_{22} & S_{23} & S_{24} \\ S_{31} & S_{32} & S_{33} & S_{34} \\ S_{41} & S_{42} & S_{43} & S_{44} \end{pmatrix} \begin{pmatrix} I_{inc} \\ Q_{inc} \\ U_{inc} \\ V_{inc} \end{pmatrix}, \tag{1.92}$$

where

$$S_{11} = \frac{1}{2}\left(|S_1|^2 + |S_2|^2 + |S_3|^2 + |S_4|^2\right), \tag{1.93}$$

gives the angular variation of the scattered intensity. Details on other elements of the Stokes matrix and their relations with ellipsometric parameters, Jones matrices and Poincare sphere representations can be found in many textbooks (see, for example, Guenther 1990).

If the scattered wave is a cylindrical wave, the scattering matrix is defined as

$$\begin{pmatrix} E_{\parallel}^s \\ E_{\perp}^s \end{pmatrix} = e^{i3\pi/4}\sqrt{\frac{2}{\pi k r \sin\zeta}}\, e^{ik(r\sin\zeta - z\cos\zeta)} \begin{pmatrix} T_1 & T_4 \\ T_3 & T_2 \end{pmatrix} \begin{pmatrix} E_{\parallel}^i \\ E_{\perp}^i \end{pmatrix}, \tag{1.94}$$

where the incident beam makes an angle ζ with the axis of the cylinder. This becomes

$$\begin{pmatrix} E_{\parallel}^s \\ E_{\perp}^s \end{pmatrix} = e^{i3\pi/4}\sqrt{\frac{2}{\pi k r}}\, e^{ikr} \begin{pmatrix} T_1 & 0 \\ 0 & T_1 \end{pmatrix} \begin{pmatrix} E_{\parallel}^i \\ E_{\perp}^i \end{pmatrix}, \tag{1.95}$$

for perpendicular incidence.

1.6 Acoustic wave scattering

Having identified a similarity between the time independent Schroedinger equation in the presence of a potential $V(\mathbf{r})$ and the scalar wave equation in the presence of a particle of refractive index $m(\mathbf{r})$, we now look at the propagation of acoustic waves in a medium characterized by compressibility κ_0 and density ρ_0. The wave equation for the propagation of acoustic pressure, $p(\mathbf{r}, t)$, can be written as:

$$\nabla^2 p(\mathbf{r}, t) - \rho_0 \kappa_0 \frac{\partial^2 p(\mathbf{r}, t)}{\partial t^2} = 0, \tag{1.96}$$

which represents a wave travelling with velocity

$$c_0 = \frac{1}{\sqrt{\kappa_0 \rho_0}}. \tag{1.97}$$

The wave impedance is

$$Z = \sqrt{\frac{\rho_0}{\kappa_0}}. \tag{1.98}$$

The velocity c_0 may be identified as the characteristic speed for acoustic waves similar to the speed c of electromagnetic waves in vacuum or the particle velocity in free space in the Schroedinger equation. For sound waves, it may be taken as its speed in air. The parameters ρ_0 and κ_0, respectively, can be seen to be analogues of μ_0 and ε_0.

The wave equation in a medium of density $\rho_e(\mathbf{r})$ and compressibility $\kappa_e(\mathbf{r})$ can be expressed as (Morse and Ingard 1968):

$$\nabla^2 p - \frac{1}{c^2}\frac{\partial^2 p}{\partial t^2} = \gamma_\kappa(\mathbf{r})\frac{1}{c^2}\frac{\partial^2 p}{\partial t^2} + \mathrm{div}\big(\gamma_\rho(\mathbf{r})\mathrm{grad}\ p\big), \tag{1.99}$$

where

$$\gamma_\kappa(\mathbf{r}) = \frac{(\kappa_e(\mathbf{r}) - \kappa_o)}{\kappa_o}, \tag{1.100}$$

and

$$\gamma_\rho(\mathbf{r}) = \frac{(\rho_e(\mathbf{r}) - \rho_o)}{\rho_o}. \tag{1.101}$$

The first term on the right-hand side in (1.99) causes scattering by variations in compressibility and the second term causes scattering by changes in density.

If the acoustic motion has a single frequency, ω, and interaction is time independent, we may write,

$$p(\mathbf{r}, t) = p_\omega e^{-i\omega t}, \tag{1.102}$$

The time independent wave equation then becomes,

$$\nabla^2 p_\omega(\mathbf{r}) + k^2 p_\omega(\mathbf{r}) = -k^2\gamma_\kappa(\mathbf{r})p_\omega(\mathbf{r}) + \nabla.\Big(\gamma_\rho(\mathbf{r})\mathrm{grad}\ p_\omega(\mathbf{r})\Big). \tag{1.103}$$

A comparison of (1.103) with (1.58) shows that the two equations are formally identical. One only needs to make the following replacements,

$$V(\mathbf{r}) = \frac{\hbar^2 k^2}{2M}\left(m^2(\mathbf{r}) - 1\right) = \frac{-\hbar^2}{2M}\left[k^2\gamma_\kappa(\mathbf{r}) - \nabla.\Big(\gamma_\rho(\mathbf{r})\mathrm{grad}\Big)\right]. \tag{1.104}$$

The scattering amplitude, denoted as $f_{ac}(\theta.\varphi)$, may then be expressed as,

$$f_{ac}(\theta, \varphi) = \frac{k^2}{4\pi}\int e^{-i\mathbf{k}_s.r}\left[\gamma_\kappa(\mathbf{r})p_\omega(\mathbf{r}) - i\gamma_\rho(\mathbf{r})\frac{\hat{\mathbf{a}}_s}{k}.\nabla p_\omega(\mathbf{r})\right], \tag{1.105}$$

where the subscript ac stands for acoustic and $\hat{\mathbf{a}}_s$ is a unit vector along the direction of the detector.

The acoustic wave scattering and the electromagnetic wave scattering are regarded as complementary techniques. Two main advantages of the acoustic (ultrasound)-based techniques over electromagnetic scattering-based techniques are: (i) ultrasound can penetrate optically opaque mixtures. Therefore, no dilution of the sample target is needed in ultrasound-based techniques. In contrast, optical techniques for opaque mixtures require dilution which can easily cause alteration in characteristic properties of the original sample. (ii) Acoustic waves have much longer wavelengths in comparison to the optical wavelengths. Hence, simple small particle approximations, such as Born approximation, can still be used in acoustic scattering for particles which have become large compared to the electromagnetic wavelengths in the visible region.

Chapter 2

Single particle scattering

2.1 Analytic solutions

Basically two approaches have been employed for obtaining analytic solutions for scattering by a single particle. The first approach divides an entire space into separate refractive index regions. The fields are written as expansions in terms of vector harmonics in each region. The coefficients of expansion are obtained by matching the solutions at the boundaries. The angular scattering functions and various efficiency factors can then be written as infinite series in terms of these coefficients. This method of obtaining a solution is called the boundary condition method. A major limitation of this approach is that it can be applied only to particles of certain regular geometrical shapes.

The second approach emanates from the integral equation formulation of the Maxwell equations. It allows scattering amplitude to be expressed as an integral whose evaluation requires the solution of the relevant wave equation within the scattering region. The prime limitation of this perspective is that this does not yield an exact solution readily. Instead, it is more suitable for achieving analytic solutions of an approximate nature. Obviously, this approach is very useful for obtaining solutions in problems where exact solutions are not available. However, the approximation methods are often preferable even when the exact solutions are available. This is because the approximate solutions are simple and generally lead to better and clearer understandings of the physical processes involved. A number of approximation methods suitable for applications in various size and refractive index regions have been developed and the interest in improved approximation methods continues unabated even today despite the availability of faster and faster computers and improved numerical methods.

A homogeneous isotropic sphere was the first object for which the problem of electromagnetic scattering was solved exactly (Lorenz 1890). But Lorenz's work did not receive due attention. This was perhaps because it was in the Danish language. Later, a solution for scattering by a homogeneous sphere was suggested by Love (1899). However, the credit for a complete solution is generally given to Gustav Mie (1908) who even computed the first few terms of the series solution. The rigorous theory of scattering by a homogeneous isotropic sphere is, therefore, widely referred to as the Mie theory. Some authors prefer to call it the Lorenz–Mie theory (see, for example, Grandy 2000, Gouesbet and Grehan 2011). The related problem of radiation pressure on a sphere was considered by Debye (1909) independently. Thus, the theory has also been assigned the name of the Lorenz–Mie–Debye theory (see, for example, Davis and Schweiger 2002). Nevertheless, the most common terminology refers to the Mie theory. More details about the Mie theory and its history can be found in many reviews and books (see, for example, Logan 1965, Hovarth 2009, Mishchenko 2009, Hergert and Wriedt 2012).

The Mie solution assumes the host medium to be nonabsorbing. The solutions are cast as infinite series involving spherical Riccati–Bessel, Riccati–Neumann and Riccati–Hankel functions. The numerical computation of these functions remained a tedious job for a long time and not much incentive was seen by researchers in investing time and energy for developing solutions for particles of other shapes. The first major numerical computation seems to be for spheres of diameters up to 3.2λ by Gans (1925) and by Blumer (1925). Apparently there is some inaccuracy in the later work that was pointed out by Logan (1965). The extensive tables of Mie functions were prepared and published by Lowan (1949). Tables of scattering functions were prepared for specified values of particle-size parameters, relative refractive indices and scattering angles by Denman, Heller and Pangonis (1966) and by Wickramasinghe (1973). A bibliography of selected scattering functions can be found in Kerker (1969).

Given the limitations of numerical computation about a century ago, it is not surprising that it took about 44 years before analytic solutions for a homogeneous sphere were extended to a concentric sphere (Aden and Kerker 1951). Güttler (1952) and Shifrin (1952) also solved the problem independently. Exact solutions for a circular cylinder by Wait (1955), for a coaxial cylinder by Adey (1956) and for radially inhomogeneous sphere by Wyatt (1962) and Wait (1963) followed soon. The statistics in an article by Shore (2015) shows that the original paper by Mie has been cited in over 6200 journal articles. Of these, about 5800 are in the past 50 years. Clearly, the Mie theory and its augmentations began influencing research in the field of electromagnetic wave scattering only after large-scale computations became feasible.

With the advent of high speed computers, it has become possible to compute scattering functions routinely even for very large particles and for various complex particle shapes. Exact analytic solutions have been extended also to include particles of the same shape but of different physical attributes. For example, the Mie solution for an isotropic sphere was extended to include optically active spheres (Bohren 1974, Bohren and Gilra 1979), charged spheres (Bohren and Hunt 1977, Klacka and Kocifaj 2007, 2010), spheres with axial anisotropy (Monzon 1989), etc. The solution was obtained for scattering by a focused Gaussian beam (Gouesbet and Grehan 1982, 2011). The resulting theory is known as the Generalized Lorenz–Mie theory (GLMT). The conventional Mie solution has also been extended to include the problem of scattering by a sphere in an absorbing host medium.

Numerical solutions constitute a very important part of the electromagnetic wave scattering solutions. A review of elastic light scattering theories and numerical methods can be found in Wriedt (1998) and Wriedt and Comeberg (1998). Mishchenko et al. (1996) has given an exhaustive review of T matrix methods. The point matching methods have been reviewed by Bates (1973). The coupled dipole method, also known as the Purcell–Pennypacker method (1973) and the discrete dipole method (Draine and Flatau 1984, Draine 1988, Yurkin and Hoekstra 2011) are other popular numerical approaches. A review of single light scattering computational methods can be found in Farafonov and Il'in (2006) also.

2.2 Rigorous analytic solutions

Two of the commonly used shapes in modelling particles are: (i) homogeneous spheres and (ii) layered spheres. Rigorous analytic solutions for the scattering of electromagnetic waves by these two shapes are considered below.

2.2.1 Homogeneous sphere: Mie scattering

Consider a plane electromagnetic wave of wavelength λ_0 (in vacuum), incident on a homogeneous sphere of radius a. The host medium is infinite, homogeneous, isotropic and nonabsorbing. The wave number of the wave in the medium is $k = 2\pi N/\lambda_0$, N being the real refractive index of the host medium. The corresponding dielectric constant and magnetic permeability are represented by ε and μ. The size parameter of the sphere is $x = ka$. Its complex refractive index, dielectric constant and magnetic permeability are denoted by m_1, ϵ_1 and μ_1 respectively. The refractive index of the sphere with respect to the surrounding medium is denoted by $m = m_1/N = n_r + in_i$.

The solution of this problem has been described in detail in many textbooks on light and/or electromagnetic scattering. Some of the widely referred books include Stratton (1941), van de Hulst (1957), Kerker (1969), Bayvel and Jones (1981), Bohren and Huffman (1983) and Mishchenko et al. (2002). The treatment adopted here closely follows that in Bohren and Huffman (1983).

Briefly, the first step towards obtaining a solution to the vector wave equations,

$$\nabla^2\mathbf{E} + k^2\mathbf{E} = 0, \tag{2.1}$$

and

$$\nabla^2\mathbf{H} + k^2\mathbf{H} = 0, \tag{2.2}$$

is to construct a set of two vector functions which themselves satisfy vector wave equations. Two vectors $\mathbf{M}$ and $\mathbf{N}$ are thus defined as,

$$\mathbf{M} = \nabla \times (\mathbf{c}\psi), \tag{2.3}$$

and

$$\mathbf{N} = \frac{\nabla \times \mathbf{M}}{k}, \tag{2.4}$$

where ψ is a scalar function and $\mathbf{c}$ is a constant vector. The vectors $\mathbf{M}$ and $\mathbf{N}$ are called vector spherical harmonics and obey the following equations,

$$\nabla.\mathbf{M} = 0, \qquad \nabla.\mathbf{N} = 0, \tag{2.5}$$

$$\nabla^2\mathbf{M} + k^2\mathbf{M} = 0. \tag{2.6}$$

$$\nabla^2\mathbf{N} + k^2\mathbf{N} = 0. \tag{2.7}$$

and

$$\nabla \times \mathbf{N} = k\mathbf{M}. \tag{2.8}$$

provided,

$$(\nabla^2 + k^2)\psi = 0, \tag{2.9}$$

The vectors $\mathbf{M}$ and $\mathbf{N}$ have all the properties of an electromagnetic field. If $\mathbf{c}$ is chosen to be the radius vector of the sphere, $\mathbf{M}$ is everywhere tangential to spherical surface.

Two independent solutions of vector Helmholtz equation (2.6) in a spherical coordinate system are known to be,

$$\psi^{(j)}_{e,m,n}(r,\theta,\phi) = z^{(j)}_n(r)P^m_n(\theta)\cos(m\phi), \tag{2.10}$$

$$\psi^{(j)}_{o,m,n}(r,\theta,\phi) = z^{(j)}_n(r)P^m_n(\theta)\sin(m\phi), \tag{2.11}$$

where P^m_n is the associated Legendre polynomial and $z_n(r)$ could be either the Bessel function $j_n(r)$, Neumann function $y_n(r)$ or a Hankel function

$$h^{(1)}_n(r) = j_n(r) + iy_n(r), \tag{2.12}$$

or

$$h^{(2)}_n(r) = j_n(r) - iy_n(r). \tag{2.13}$$

Any function satisfying the scalar wave equation in spherical coordinates can be expanded as an infinite series in terms of functions ψ_{emn} and ψ_{omn}. Vector spherical harmonics (2.3) and (2.4) can therefore be written as,

$$\mathbf{M}_{emn} = \nabla \times (\mathbf{r}\psi_{emn}), \quad \mathbf{M}_{omn} = \nabla \times (\mathbf{r}\psi_{emn}), \tag{2.14}$$

and

$$\mathbf{N}_{emn} = \frac{\nabla \times \mathbf{M}_{emn}}{k}, \quad \mathbf{N}_{omn} = \frac{\nabla \times \mathbf{M}_{omn}}{k}. \tag{2.15}$$

The desired expansion of electric and magnetic vectors is then,

$$\mathbf{E}_i = E_n \sum_{n=1}^{n=\infty} \left(\mathbf{M}^{(1)}_{o1n} - i\mathbf{N}^{(1)}_{e1n}\right), \tag{2.16}$$

and

$$\mathbf{H}_i = E_n \sum_{n=1}^{n=\infty} \left(\mathbf{M}^{(1)}_{e1n} + i\mathbf{N}^{(1)}_{o1n}\right), \tag{2.17}$$

where

$$E_n = i^n E_0 \frac{2n+1}{n(n+1)}. \tag{2.18}$$

The superscript (1) in (2.16) and (2.17) implies vector spherical harmonics for which the radial dependence of generating functions is specified by j_n.

The expansion of the field inside the sphere is

$$\mathbf{E}_1 = E_n \sum_{n=1}^{n=\infty} \left(c_n\mathbf{M}^{(1)}_{o1n} - id_n\mathbf{N}^{(1)}_{e1n}\right), \tag{2.19}$$

and

$$\mathbf{H}_1 = \frac{k_1}{\omega\mu} E_n \sum_{n=1}^{n=\infty} \left(d_n\mathbf{M}^{(1)}_{e1n} + ia_n\mathbf{N}^{(1)}_{o1n}\right), \tag{2.20}$$

where k_1 is the wave number associated with the wave inside the sphere. Similarly, the expansion of the scattered field can be written as

$$\mathbf{E}_s = \sum_{n=1}^{n=N} E_n \left(ia_n \mathbf{N}_{e1n}^{(3)} - b_n \mathbf{M}_{o1n}^{(3)} \right), \tag{2.21}$$

and

$$\mathbf{H}_s = \sum_{n=1}^{n=N} E_n \left(ib_n \mathbf{N}_{o1n}^{(3)} + a_n \mathbf{M}_{e1n}^{(3)} \right). \tag{2.22}$$

The superscript (3) in (2.21) and (2.22) implies vector spherical harmonics specified by the Hankel function $h_n^{(1)}$.

The application of the boundary condition,

$$(\mathbf{E}_i + \mathbf{E}_s - \mathbf{E}_1) \times \hat{\mathbf{e}}_r = (\mathbf{H}_i + \mathbf{H}_s - \mathbf{H}_1) \times \hat{\mathbf{e}}_r = 0, \tag{2.23}$$

leads to four linear equations involving four unknown coefficients. When solved, these give:

$$a_n = \frac{\mu m \psi_n(mx) \psi_n'(x) - \mu_1 \psi_n(x) \psi_n'(mx)}{\mu m \psi_n(mx) \xi_n'(x) - \mu_1 \xi_n(x) \psi_n'(mx)}, \tag{2.24}$$

$$b_n = \frac{\mu_1 \psi_n(mx) \psi_n'(x) - \mu m \psi_n(x) \psi_n'(mx)}{\mu_1 \psi_n(mx) \xi_n'(x) - \mu m \xi_n(x) \psi_n'(mx)}, \tag{2.25}$$

$$c_n = \frac{\mu_1 \psi_n(x) \xi_n'(x) - \mu_1 \xi_n(x) \psi_n'(x)}{\mu_1 \psi_n(mx) \xi_n'(x) - \mu m \psi_n(x) \xi_n'(mx)}, \tag{2.26}$$

and

$$d_n = \frac{\mu_1 m \psi_n(x) \xi_n'(x) - \mu_1 m \xi_n(x) \psi_n'(x)}{\mu m \psi_n(mx) \xi_n'(x) - \mu_1 \xi_n(x) \psi_n'(mx)}. \tag{2.27}$$

The primes denote differentiation with respect to the argument. As before, $\xi_n(x) = x h_n^{(1)}(x)$ and $\psi_n(x) = x j_n(x)$ are the Riccati–Hankel and Riccati–Bessel functions, respectively.

In a non-absorbing host medium, only the coefficients a_n and b_n are of any significance. Further, for optical wavelengths $\mu = \mu_1$ in most practical situations. Thus, a_n and b_n become

$$a_n = \frac{m \psi_n(mx) \psi_n'(x) - \psi_n(x) \psi_n'(mx)}{m \psi_n(mx) \xi_n'(x) - \xi_n(x) \psi_n'(mx)} \equiv \frac{A_n(m,x)}{C_n(m,x)}, \tag{2.28}$$

$$b_n = \frac{\psi_n(mx) \psi_n'(x) - m \psi_n(x) \psi_n'(mx)}{\psi_n(mx) \xi_n'(x) - m \xi_n(x) \psi_n'(mx)} \equiv \frac{B_n(m,x)}{D_n(m,x)}. \tag{2.29}$$

An alternative way of writing Mie coefficients a_n and b_n is,

$$a_n = \frac{1 - e^{2i\alpha_n}}{2} = -ie^{i\alpha_n} \sin \alpha_n, \tag{2.30}$$

$$b_n = \frac{1 - e^{2i\beta_n}}{2} = -ie^{i\beta_n} \sin \beta_n, \tag{2.31}$$

where

$$\tan \alpha_n = \frac{m\psi_n(mx)\psi_n'(x) - \psi_n(x)\psi_n'(mx)}{-m\psi_n(mx)v_n'(x) + v_n(x)\psi_n'(mx)}, \tag{2.32}$$

and

$$\tan \beta_n = \frac{\psi_n(mx)\psi_n'(x) - m\psi_n(x)\psi_n'(mx)}{-\psi_n(mx)v_n'(x) + mv_n(x)\psi_n'(mx)}, \tag{2.33}$$

with $v_n = -xy_n(x)$ as the Riccati–Neumann function. The real and the imaginary parts of a_n are:

$$\mathrm{Re}\,[a_n] = \sin^2 \alpha_n; \qquad \mathrm{Im}\,[a_n] = -\cos \alpha_n \sin \alpha_n. \tag{2.34}$$

Expressions similar to (2.34) hold for b_n.

The angular scattering amplitudes $S_1(\theta)$ and $S_2(\theta)$ are given by the expressions

$$S_1(\theta) = \sum_{n=1}^{\infty} \frac{2n+1}{n(n+1)} \Big[a_n \pi_n(\cos\theta) + b_n \tau_n(\cos\theta) \Big], \tag{2.35}$$

$$S_2(\theta) = \sum_{n=1}^{\infty} \frac{2n+1}{n(n+1)} \Big[a_n \tau_n(\cos\theta) + b_n \pi_n(\cos\theta) \Big], \tag{2.36}$$

where

$$\pi_n(\cos\theta) = \frac{P_n^1(\cos\theta)}{\sin\theta}, \quad \tau_n(\cos\theta) = \frac{d}{d\theta} P_n^1(\cos\theta), \;\; P_n^1(\cos\theta) = -\frac{d}{d\theta} P_n(\cos\theta), \tag{2.37}$$

with $P_n(\cos\theta)$ as the Legendre polynomial of degree n. The amplitudes S_3 and S_4 vanish for a spherically symmetric obstacle.

If the incident wave is polarized perpendicular to the scattering plane,

$$i_\perp(\theta) = |S_1(\theta)|^2. \tag{2.38}$$

If the incident wave is polarized parallel to the scattering plane,

$$i_\parallel(\theta) = |S_2(\theta)|^2. \tag{2.39}$$

For a polarized wave with E vector making an angle ϕ_p with the normal to the scattering plane,

$$i_{sca}(\theta) = \left(i_\parallel \sin^2 \phi_p + i_\perp \cos^2 \phi_p \right). \tag{2.40}$$

If the wave is unpolarized, the scattered field intensities perpendicular or parallel to the plane of scattering are found by averaging over ϕ_p. The averaging of $\cos^2 \phi_p$ and $\sin^2 \phi_p$ over $\phi_p = 0$ to 2π is known to yield 1/2. Thus,

$$i_{sca}(\theta) = \frac{(i_{\parallel} + i_{\perp})}{2}. \tag{2.41}$$

The polarization of the scattered wave is defined as,

$$P = \frac{i_{\perp} - i_{\parallel}}{i_{\perp} + i_{\parallel}}, \tag{2.42}$$

and its absolute value is the degree of polarization.

For forward scattering ($\theta = 0$), both polarizations lead to the same result,

$$S_1(0) = S_2(0) = \frac{1}{2}\sum_{n=1}^{\infty}(2n+1)\left[a_n + b_n\right]. \tag{2.43}$$

This may be contrasted with the corresponding expression for the scalar wave scattering,

$$S(0) = \sum_{n=0}^{\infty}(2n+1)b_n, \tag{2.44}$$

Note that while the summation in (2.44) is from $n = 0$ to ∞, it is from $n = 1$ to ∞ in (2.43).

Another quantity of interest, particularly from the point of view of radiative transfer studies, is the phase function. With explicit parameter dependence exhibited, it can be represented as,

$$\phi(m, k, a, \theta) = \frac{1}{k^2}\frac{i(m, x, \theta)}{C_{sca}(m, k, a)}, \tag{2.45}$$

where $C_{sca}(m, k, a)$ is the total scattering cross section of the sphere, defined as,

$$C_{sca}(m, k, a) = \frac{2\pi}{k^2}\int_0^{\pi} i(m, x, \theta)\sin\theta \; d\theta, \tag{2.46}$$

and the phase function is normalized as,

$$2\pi\int_0^{\pi} \phi(m, x, \theta, \phi)\sin\theta \; d\theta = 1. \tag{2.47}$$

This permits the phase function to be interpreted as giving the probability of radiation scattering at an angle θ relative to the incident direction.

In Chapter 1, the efficiency factors for scalar wave scattering were defined via equation (1.66). The same definitions remains valid for scattering by electromagnetic waves. However, a more appropriate and widely applied elucidation for efficiency factors in this context is:

$$Q_{abs} = \frac{W_{abs}}{f}, \quad Q_{sca} = \frac{W_{sca}}{f}, \quad Q_{ext} = \frac{W_{ext}}{f}, \tag{2.48}$$

where $f = \pi a^2 I_i$, I_i is the incident irradiance. W_{abs} and W_{sca}, respectively, denote the rates at which the energy is absorbed and scattered. W_{ext} is the sum of the two. That is,

$$W_{ext} = W_{abs} + W_{sca}. \tag{2.49}$$

Denoting the total electric and magnetic fields $\mathbf{E}$ and $\mathbf{H}$ at a point outside the sphere as the sum of the incident and scattered fields:

$$\mathbf{E} = \mathbf{E}_{inc} + \mathbf{E}_{sca}, \tag{2.50}$$

and

$$\mathbf{H} = \mathbf{H}_{inc} + \mathbf{H}_{sca}, \tag{2.51}$$

it is straightforward to see that W_{sca} and W_{ext} can be expressed as,

$$W_{sca} = \int_A \mathbf{S}_{sca}.\hat{\mathbf{e}}_r \, dA, \qquad W_{ext} = \int_A \mathbf{S}_{ext}.\hat{\mathbf{e}}_r \, dA, \tag{2.52}$$

with relevant Poynting vectors given by:

$$\mathbf{S}_{sca} = \frac{1}{2}\text{Re}\left[\mathbf{E}_{sca} \times \mathbf{H}^*_{sca}\right], \tag{2.53}$$

and

$$\mathbf{S}_{ext} = \frac{1}{2}\text{Re}\left[\mathbf{E}_{inc} \times \mathbf{H}^*_{sca} + \mathbf{E}_{sca} \times \mathbf{H}^*_{inc}\right]. \tag{2.54}$$

The integration is over the surface A of an imaginary sphere which encloses the scatterer, and $\hat{\mathbf{e}}_r$ is the outward unit vector normal to the surface of the sphere. The asterisk denotes the complex conjugate of the field. The efficiency factors for extinction, scattering and absorption, respectively, can be expressed in terms of Mie coefficients as,

$$Q_{ext} = \frac{2}{x^2}\sum_{n=1}^{\infty}\Big[(2n+1)\text{Re}(a_n + b_n)\Big], \tag{2.55}$$

$$Q_{sca} = \frac{2}{x^2}\sum_{n=1}^{\infty}\Big[(2n+1)(|a_n|^2 + |b_n|^2)\Big], \tag{2.56}$$

and

$$Q_{abs} = Q_{ext} - Q_{sca} = \frac{2}{x^2}\sum_{n=1}^{\infty}(2n+1)\left[\frac{1}{2} - \left|a_n - \frac{1}{2}\right|^2 - \left|b_n - \frac{1}{2}\right|^2\right], \tag{2.57}$$

For a nonabsorbing scatterer $Q_{ext} = Q_{sca}$. It follows from (2.55) and (2.56) that

$$\text{Re } a_n = |a_n|^2, \tag{2.58}$$

and

$$\text{Re } b_n = |b_n|^2. \tag{2.59}$$

Further, since $Q_{abs} = 0$,

$$\left|a_n - \frac{1}{2}\right| = \left|b_n - \frac{1}{2}\right| = \frac{1}{2}. \tag{2.60}$$

That is, the Mie coefficients lie on a circle in a complex plane with the origin given by $(1/2, 0)$. The radius of the sphere is $1/2$.

The asymmetry parameter g can also be stated as an analytic expression for the present problem:

$$g = \frac{4}{x^2}\left[\sum_{n=1}^{\infty} \frac{n(n+2)}{n+1} \text{Re}\left(a_n a_{n+1}^* + b_n b_{n+1}^*\right) + \sum_{n=1}^{\infty} \frac{2n+1}{n(n+1)} \text{Re}\left(a_n b_n^*\right)\right], \tag{2.61}$$

in terms of a_n and b_n.

The extinction efficiency factor for backscattering is defined as,

$$Q_{back} = \frac{4}{x^2}\, i(180^o). \tag{2.62}$$

Since,

$$S_2(180) = -S_1(180) = \frac{1}{2}\sum_{n=1}^{\infty}(2n+1)(-1)^n(a_n - b_n), \tag{2.63}$$

the efficiency factor for backscattering by a sphere can be formulated as,

$$Q_{back} = \frac{1}{x^2}\left|\sum_{n=1}^{\infty}(2n+1)(-1)^n(a_n - b_n)\right|^2, \tag{2.64}$$

in terms of Mie scattering coefficients.

Many applications of Mie scattering require evaluation of an integral of type,

$$\int_{w_1}^{w_2} \left[|S_1(\theta)|^2 + |S_2(\theta)|^2\right] d\mu, \tag{2.65}$$

between two given angles w_1 and w_2. This integral can be evaluated analytically for scattering in a backward hemisphere (Chylek 1973). This was accomplished by defining two new functions,

$$F_+(\theta) = \frac{1}{\sqrt{2}}\left[S_1(\theta) + S_2(\theta)\right], \tag{2.66}$$

and

$$F_-(\theta) = \frac{1}{\sqrt{2}} \left[S_1(\theta) - S_2(\theta) \right], \tag{2.67}$$

such that,

$$\int \left[|S_1(\theta)|^2 + |S_2(\theta)|^2 \right] d\mu = \int \left[|F_+(\theta)|^2 + |F_-(\theta)|^2 \right] d\mu. \tag{2.68}$$

Substituting for $S_1(\theta)$ and $S_2(\theta)$ from (2.35) and (2.36), the right-hand side of (2.68) can be couched as,

$$\int_{w_1}^{w_2} \left[|F_+(\theta)|^2 + |F_-(\theta)|^2 \right] d\mu \equiv I_1 + I_2 + I_3, \tag{2.69}$$

where

$$I_1 = \sum_{k=1}^{\infty} \sum_{l=1}^{\infty} (2k+1)(2l+1)(a_k a_l^* + b_k b_l^*) I_{lk}(w_1, w_2), \tag{2.70}$$

$$I_2 = \sum_{k=1}^{\infty} \sum_{l=1}^{\infty} \frac{(2k+1)(2l+1)}{k(k+1)(l+1)} (a_k a_l^* + b_k b_l^*) \left[\mu(1-\mu^2) \pi_k(\mu) \pi_l(\mu) \right]_{\mu=w_1}^{\mu=w_2}, \tag{2.71}$$

$$I_3 = -\sum_{k=1}^{\infty} \sum_{l=1}^{\infty} \frac{(2k+1)(2l+1)}{k(k+1)(l+1)} (a_k b_l^* + b_k a_l^*) \left[\mu(1-\mu^2) \pi_k(\mu) \pi_l(\mu) \right]_{\mu=w_1}^{\mu=w_2}, \tag{2.72}$$

with

$$I_{lk}(w_1, w_2) = \int_{w_1}^{w_2} P_l(\mu) P_k(\mu) \, d\mu. \tag{2.73}$$

The term I_{lk}, appearing in (2.70), proved to be a stumbling block in getting an analytic formula for arbitrary w_1 and w_2. However, an analytic expression could be secured for the particular case of a backward hemisphere, that is, when $w_1 = -1$ and $w_2 = 0$. Rearranging the terms after evaluation of (2.73), a compact form that emerges is,

$$\begin{aligned} \int \left[|S_1(\theta)|^2 + |S_2(\theta)|^2 \right] d\mu &= \sum_{k=1}^{\infty} (2k+1) \left(|a_k|^2 + |b_k|^2 \right) \\ &+ 2 \sum_{k=2}^{\infty}{}'' \sum_{l=1}^{\infty}{}' (-1)^{(k+l-1)/2} \frac{(2k+1)(2l+1)}{(k-l)(k+l+1)} \frac{(k-1)!!l!!}{k!!(l-1)!!} \mathrm{Re} \left(a_k a_l^* + b_k b_l^* \right) \\ &+ 2 \sum_{k=1}^{\infty}{}' \sum_{l=1}^{\infty}{}' (-1)^{(k+l)/2} \frac{(2k+1)(2l+1)}{k(k+1)(l-1)} \frac{k!!l!!}{(k-1)!!(l-1)!!} \mathrm{Re} \left(a_k b_l^* \right). \end{aligned} \tag{2.74}$$

where $\sum'$ and $\sum''$ denote summation over odd and even numbers, respectively. Computations from (2.74) have been contrasted with Mie theory results by Chylek et al. (1975). The agreement between the two is very good indeed.

An analytic expression for the above problem was obtained for arbitrary w_1 and w_2 by Wiscombe and Chylek (1977). It was discovered that when $l = k$, a closed form expression for the integral I_{lk} already existed in the literature (Erdelyi 1953). For $l \neq k$, a recursion relation was developed successfully for computing I_{lk}.

2.2.2 Mie theory in Gegenbauer polynomials

The Mie scattering functions can be expressed in terms of Gegenbauer polynomials too (Ludlow and Everitt 1995, 1996). By writing,

$$\pi_n(\cos\theta) = \frac{P_n^1(\cos\theta)}{\sin\theta} = T_{n-1}^1(\cos\theta), \tag{2.75}$$

and

$$\tau_n(\cos\theta) = \frac{d}{d\theta}P_n^1(\cos\theta) = \frac{1}{2n+1}\left[n^2 T_n^1(\cos\theta) - (n+1)^2 T_{n-2}^1(\cos\theta)\right], \tag{2.76}$$

where T_n^1 is the Gegenbauer polynomial of order n and degree 1, the scattering functions $S_1(\theta)$ and $S_2(\theta)$ given in (2.35) and (2.36) can be cast as,

$$S_1(\theta) = \sum_{n=0}^{n=\infty} C_n^{(1)} T_n^1(\cos\theta), \tag{2.77}$$

and

$$S_2(\theta) = \sum_{n=0}^{n=\infty} C_n^{(2)} T_n^1(\cos\theta), \tag{2.78}$$

where

$$C_n^{(1)} = \frac{n}{n+1}b_n + \frac{2n+3}{(n+1)(n+2)}a_{n+1} - \frac{n+3}{n+2}b_{n+2}, \tag{2.79}$$

and

$$C_n^{(2)} = \frac{n}{n+1}a_n + \frac{2n+3}{(n+1)(n+2)}b_{n+1} - \frac{n+3}{n+2}a_{n+2}, \tag{2.80}$$

Enunciating the scattering functions in terms of Gegenbauer polynomials has the following advantages. (i) As will be seen in the next chapter, it allows a closer comparison of the Mie theory with the Rayleigh–Gans approximation. (ii) For numerical computations, the convergence of the series seems to be faster in comparison to the convergence in the conventional Mie series. (iii) This form appears to have some advantage in solving the inverse scattering problems too (Ludlow and Everitt 2000).

2.2.3 Computation of Mie coefficients

A study of intensity curves as a function of $\mathcal{N}$, the number of terms needed in the summation, shows that the convergence is oscillatory and quite irregular (Bryant and Cox 1966). Thus, although the Mie coefficients a_n and b_n approach zero as $n \to \infty$, it is necessary to know the number of terms, $n = \mathcal{N}$, beyond which the summation may be terminated. Wiscombe (1980) arrived at the following criterion:

$$\mathcal{N} = \text{Int}\left(x + 4.05x^{1/3}\right) + 2.0. \tag{2.81}$$

Int(.) in (2.81) indicates the integer part of the number in the bracket. The condition (2.81) has been arrived at by imposing the requirement:

$$|a_n|^2 + |b_n|^2 \leq 5 \times 10^{-14}. \tag{2.82}$$

This is known as the Dave criterion. Mishchenko et al. (2002) give a criterion similar to (2.81) but with the last term as 8 instead of 2.

The dependence of $\mathcal{N}$ on m and x has been investigated by Stübinger et al. (2010) using the arbitrary precision calculator (Bell 2003). It was shown that a modified Dave criterion,

$$|a_n|^2 + |b_n|^2 \leq 5 \times 10^{-80}, \tag{2.83}$$

can be achieved without much increase in $\mathcal{N}$.

An equation with the same functional form as (2.81) but which allows flexibility in setting the amount of truncation error ϵ, has been obtained by Neves and Pisignano (2012):

$$\mathcal{N} \approx x + \alpha\epsilon^{2/3}\, x^{1/3} + \beta. \tag{2.84}$$

The truncation error ϵ is,

$$10^{-\epsilon} = \frac{\Delta Q_{sca}}{Q_{sca}} = 1 - \frac{1}{2x^2 Q_{sca}} \sum_{n=1}^{n=N} (2n+1)\left(|a_n|^2 + |b_n|^2\right), \tag{2.85}$$

and α and β are numerical constants to be determined from simulations. For the range of parameters, $1 \leq x \leq 200$ and $\epsilon < 50$, the value of $\mathcal{N}$ obtained is,

$$\mathcal{N} = x + 0.76\epsilon^{2/3} x^{1/3} - 4.1. \tag{2.86}$$

For a nonabsorbing sphere, arguments of Riccati–Bessel functions are real. The computation of scattering coefficients is then straightforward. But, for absorbing spheres, Mie coefficients may become unstable. Special tricks have been developed for circumventing this poor conditioning (Aden 1951, Kattawar and

Plass 1967, Lentz 1976, Cantrell 1988, Cachorro and Salcedo 2001). Detailed discussions on numerical computational issues can also be found in Deirmendjian (1969), Bayvel and Jones (1981) and Bohren and Huffman (1983) among many others. A recent review of computational aspects can be found in an article by Shore (2015). In addition, a comprehensive review of computational methods for single particle scattering appears in Farafonov and Il'in (2006). Numerical methods for nonspherical particles and aggregates have been summarized by Jacquier and Gruy (2010).

2.2.4 Basic structures in Mie scattering

Figure 2.1 compares the normalized angular scattering patterns for $x =$ 2.0, 10.0 and 25.0. The relative refractive index of the scatterer is $m = 1.5$ in all the plots. It may be noted that as the size of the sphere increases, the forward scattering lobe shrinks and the separation between extrema positions decreases. As a result, the slope of the forward scattering lobe, positions of the extrema and the separation between them have been related to the size of the sphere. The relationships so obtained are based either on some approximation method, such as the diffraction approximation, the anomalous diffraction approximation, the Rayleigh–Gans approximation or are a result of empirical observations.

Steiner et al. (1999) have studied the periodicity of oscillations in the scattering pattern and arrived at the following simple relation:

$$\nu = 0.00483x, \tag{2.87}$$

ν being the number of oscillations per degree. The relationship is valid when the size parameter is within the range $50 \leq x \leq 500$ and the refractive index is within the range $1.3 \leq m \leq 1.75$.

A number of empirical formulas, relating the angular separations of the first and the j-th minimum with particle size, have been obtained in the $m - x$ domain $0.9 \leq d \leq 15.0$ μm and $1.37 \leq m_1 \leq 1.60$ by Maltsev and coworkers (see, for example, Maltsev 2000) and by Shepelevich et al. (1999). The $m - x$ range above, corresponds primarily to biological particles. The refractive index of the host medium is $N = 1.333$ and the wavelength of the incident radiation (in air) is 632.8 nm. The entire $m - x$ region of interest has been divided in subregions and separate formulas have been designed for separate $m - x$ regions using a best fit procedure. The extrema number is counted after a conveniently chosen boundary angle ϕ_d. The separation between the first and the j-th minimum after ϕ_d is denoted by $\Delta\theta_j(\phi_d)$.

For $j = 3$, $\phi_d = 20$, $1.05 \leq m \leq 1.15$, and $1 \leq d \leq 12$ μm, d as given in Shepelevich et al. (1999) is,

$$d = C_1 + C_2[\Delta\theta_3(20)]^{-2} + C_3[\Delta\theta_3(20)]^{-3} + C_4[\Delta\theta_3(20)]^{-4}, \tag{2.88}$$

where $C_1 = 0.127$, $C_2 = 52$, $C_3 = 190$ and $C_4 = -660$. The argument of θ is in degrees.

Another parametrization was developed in the domain $1.125 \le m \le 1.20$, $1.0 \le d \le 10.6$, $N = 1.333$ and $\lambda_0 = 632.8\ nm$ by Chernyshev et al. (1995):

$$d = C_1 + C_2[\Delta\theta_2(20)]^{-2} + C_3[\Delta\theta_2(20)]^{-3} + C_4[\Delta\theta_2(20)]^{-4}, \tag{2.89}$$

where $C_1 = -0.6$, $C_2 = 28.8$, $C_3 = 600$ and $C_4 = -1700$. The description is a good approximation for latex particles in water.

A similar approach with $j = 2$ and a boundary angle of 15 deg gives,

$$d = C_1 + C_2[\Delta\theta_2(15)]^{-1} + C_3[\Delta\theta_2(15)]^{-5} + C_4[\Delta\theta_2(215)]^{-6}, \tag{2.90}$$

where $C_1 = -0.080$, $C_2 = 26.86$, $C_3 = 200$ and $C_4 = -420$. The domain of refractive index is $1.028 \le m \le 1.105$ and $1 \le d \le 12\ \mu m$. The $m - x$ domain of these equations corresponds to biological particles.

A relatively simpler approximation over a wider $m - x$ domain is (Maltsev 2000):

$$d \sim \frac{\lambda}{\Delta\theta_2(15)}. \tag{2.91}$$

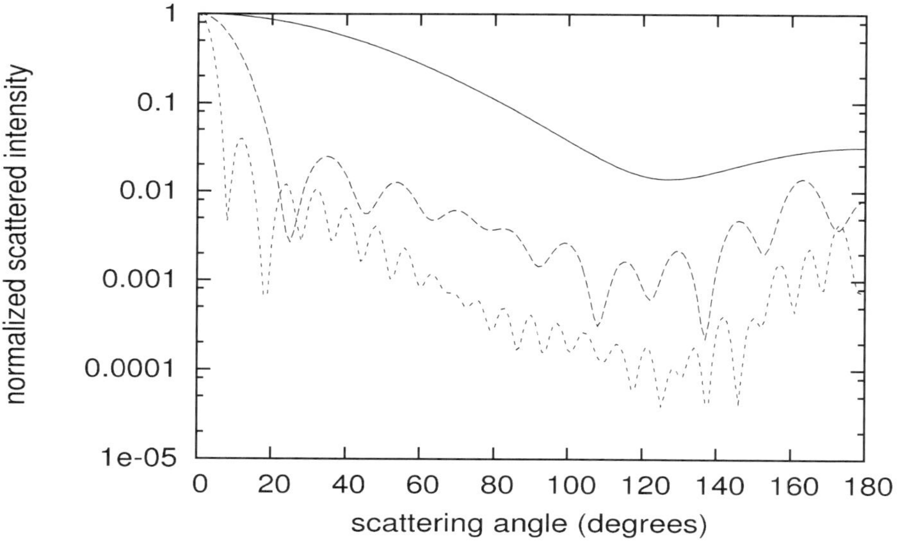

FIGURE 2.1: Normalized angular scattering patterns. Plotted for $x = 2.0$, 10.0 and 25.0. Normalization is $i(\theta) = i(\theta)/i(0)$. The solid line is for $x = 2.0$, The dashed line is for $x = 10.0$ and the small dashed line is for $x = 25.0$. The relative refractive index of the sphere is $m = 1.5$ in each case.

This connection between extrema separation and d is a good approximation in the domain $4 \leq x \leq 100$ and $1.028 \leq m \leq 1.238$. The accuracy of the approximation is of the order of λ. A similar formula:

$$d \sim \frac{54.4}{\Delta\theta_3(20)}, \tag{2.92}$$

has been given by Shepelevich et al. (1999). The $m - x$ domain for its validity is the same as that of (2.89).

The empirical formula for the relative refractive index m has also been proposed (Chernyshev et al. 1995). Towards this end, a quantity "visibility" is defined as

$$V(\phi_1) = \frac{I_{max} - I_{min}}{I_{max} + I_{min}}, \tag{2.93}$$

where I_{min} is the intensity at the angle of the next minimum after the boundary angle ϕ_1, and I_{max} is the first maximum after the boundary angle. The relationship between m and the visibility is governed by the equation,

$$m = \frac{1}{N}\left[C_1 + C_2[V(30)]^{1/2} + C_3[V(30)] + C_4[V(30)]^2\right], \tag{2.94}$$

where $C_1 = 3.9$, $C_2 = -10$, $C_3 = 12$ and $C_4 = -5$. Parameterizations for x and $\rho = 2x(m-1)$ have been obtained as (Maltsev and Lopatin 1997),

$$x = C_1 + \frac{C - 2\,[1 + C_3[V(15)]]}{\Delta_2(15)} + \frac{C_4}{[\Delta_2(15)]^3} + C_5[V(15)]^2 + C_6[V(15)]^4, \tag{2.95}$$

where $C_1 = 0.2593$, $C_2 = 178.76272$, $C_3 = -0.06009$, $C_4 = 503.43215$, $C_5 = 1.5301$, $C_6 = -0.62525$. and

$$\rho = C_1 + C_2\left[\cos^{-1}\left(C_3 + C_4\left(1 + C_5\Delta_2(15)\right)V(15) + C_6[V(15)]^2\right)\right]^2, \tag{2.96}$$

where $C_1 = 0.43432$, $C_2 = 0.39453$, $C_3 = -1.3977$, $C_4 = 2.56421$, $C_5 = 0.00858$, and $C_6 = -0.75401$. Seven different sets of coefficients have been given in Maltsev and Lopatin (1997). The formulas were studied for $N = 1.33$, $1 \leq d \leq 13.3\ \mu m$ and $1.028 \leq m \leq 1.20$.

Figure 2.2 shows a plot of scattered intensity as a function of $qa = 2ka\sin(\theta/2)$ for $x = 100.0$. It is interesting to note from the plot that the decrease in intensity exhibits a power law structure. Empirical studies reveal the following behaviour of the power law in different qa regions (Sorensen 2013),

$$\begin{aligned} &(qa)^0 \quad if \quad 0 \leq qa \leq 1, \\ &(qa)^{-2} \quad if \quad 1 \leq qa \leq 1.2\rho, \\ &(qa)^{-4} \quad if \quad qa \geq 1.2\rho. \end{aligned} \tag{2.97}$$

The spacing between successive maxima and minima in the power law regime $qa \geq 1$ is found to be:

$$\Delta\theta = \pi \qquad \text{when} \qquad \rho < 5, \tag{2.98}$$

and

$$\Delta(qa) = \frac{\pi}{ka} \qquad \text{when} \qquad \rho > 5, \tag{2.99}$$

where $\rho = 2ka|m-1|$. A considerable amount of mathematical analysis is needed to understand the intricacies of these empirical observations. Such studies have been conducted in great detail by Sorensen and coworkers (Berg 2012, Sorensen 2013 and references therein) who successfully relate the form of the pattern to the curvature of the scattering particle.

Another important structure in the Mie theory is the extinction efficiency factor Q_{ext}. Figure 2.3 shows typical variations of the Q_{ext} against the size parameter x for a homogeneous sphere. Plots for three different relative refractive indices have been shown. These are $m = 1.10$, $m = 1.50$, and $m = 1.50 + i0.1$. It can be seen that the extinction efficiency first rises almost linearly. This behaviour is similar in all three cases. Subsequently, the curves show two types of oscillations. A low frequency oscillation around the value $Q_{ext} = 2.0$, which is the asymptotic value of Q_{ext} as $x \to \infty$. For $|m-1| \to 0$, the position of the first maximum occurs at $\rho = 4.08$. Successive maxima occur at $\rho = 10.79$

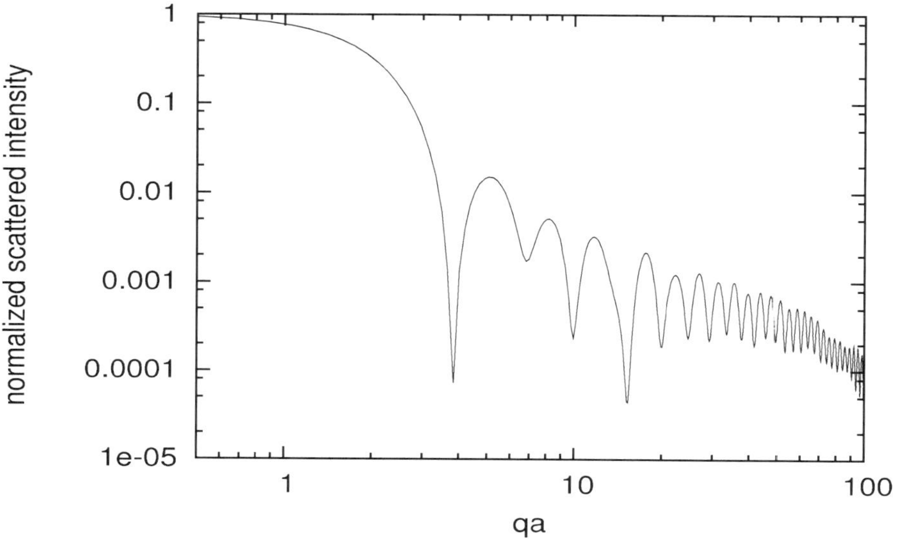

FIGURE 2.2: Normalized scattered intensity plotted as a function of qa for $x = 100.0$. The relative refractive index of the sphere is $m = 1.5$. Normalization is with respect to scattered intensity at $\theta = 0\,\text{deg}$.

and 17.16. The m dependence of the positions of maxima given by Penndorf (1958) is,

$$\rho_y(m) = \rho_y(1) + (m - 0.3), \tag{2.100}$$

where $\rho_y(m)$ is the value of ρ at the y-th maximum and $\rho_y(1)$ is the value of ρ at the y-th maximum when m is close to unity. Superimposed on this low frequency oscillation is a ripple structure (see the curve for $m = 1.5$ in Figure 2.3). While the low frequency oscillations are a result of interference between the diffracted and transmitted waves, the superimposed ripple structure is a consequence of morphology dependent resonances. For the real refractive index, the ripple structure dies down as $(m - 1)$ decreases. This is evident from the curve for $m = 1.10$ in Figure 2.3. The ripple structure also dies down as the imaginary part of the relative refractive index increases. With increasing absorption, the transmitted wave becomes less significant resulting in a weaker interference structure. The ripple structure is not of much interest in problems relating to to scattering by a collection of particles because of the averaging effect. Figure 2.3 is suggestive that the most sensitive region for extracting information about the scatterer is the monotonic rising region, but as has been shown by Roy and Sharma (2005) the other regions are also capable of yielding important information about the scatterer.

An empirical expression for the extinction efficiency factor minus a ripple obtained by van de Hulst (1957) is

$$Q_{ext} = \mathrm{Re}\left[2 - \frac{8im^2 e^{2i(m-1)x}}{x(m+1)(m^2-1)} + 4(0.46 - 0.80i)x^{-1/3}\right], \tag{2.101}$$

where $1.2 \leq m \leq 1.6$ and $x > 5$. The first term on the right-hand side of (2.101) is the contribution arising from the diffraction, the second term is the contribution of the reflection and the third term is the edge term (see, the Section 3.7.2). The applicability of the formula (2.101) has been examined by Shipley and Wienman (1978) in the range $500 < x < 4520$ and $m = 1.333$. The resulting curves showed clear agreement with the exact pattern minus the ripple structure.

An empirical formula for a large dielectric sphere is given by Walstra (1964). The result is

$$Q_{ext} = 2 - \frac{16m^2 \sin\rho}{(m+1)^2\rho} + 4\frac{1 - m\cos\rho}{\rho^2} + 7.53\frac{\bar{z} - m}{\bar{z} + m}x^{-0.772}, \tag{2.102}$$

where

$$\bar{z} = \left[(m^2 - 1)\frac{6}{\pi}^{2/3} + 1\right]. \tag{2.103}$$

This relation gives the extinction correct to within 1% for $\rho > 2.4$ and $1.0 < m \leq 1.25$. The formula is useful even for higher values of m, provided

$$x > \frac{\pi}{4(\cot^{-1} m)^3}. \tag{2.104}$$

A semi-empirical formula for $S(0)$ was also obtained by Walstra (1964):

$$S(0) = \frac{1}{2x^2} - i\frac{2xm^2 \exp(-i\rho)}{(m+1)^2(m-1)} + \frac{1 - m\exp(-i\rho)}{4(m-1)^2}$$
$$+(1.88 - i1.05)\frac{\bar{z} - m}{\bar{z} + m}x^{1.228}. \quad (2.105)$$

It is expected to reproduce a value correct to within 1% for $\rho > 3$ and $1 < m \leq 1.25$.

Penndorf (1958) has examined the variation of scattering efficiency factor for spheres in the refractive index range $1.33 \leq m \leq 1.5$. The following expressions for the value of the extinction efficiency factor at the y-th maximum:

$$Q_y(max) = 2 + \frac{4}{\rho_y(1)} + \frac{4}{\rho_y^2(1)} + \frac{29M}{\rho_y(1)} - \frac{51M}{\rho_y^2(1)}, \quad (2.106)$$

and y-th minimum:

$$Q_y(min) = 2 - \frac{4}{\rho_y(1)} + \frac{4}{\rho_y^2(1)} + \frac{8.01M}{\rho_y(1)} - \frac{27.3M}{\rho_y^2(1)}, \quad (2.107)$$

were given by him. In (2.106) and (2.107), $M = (m^2 - 1)/(m^2 + 2)$.

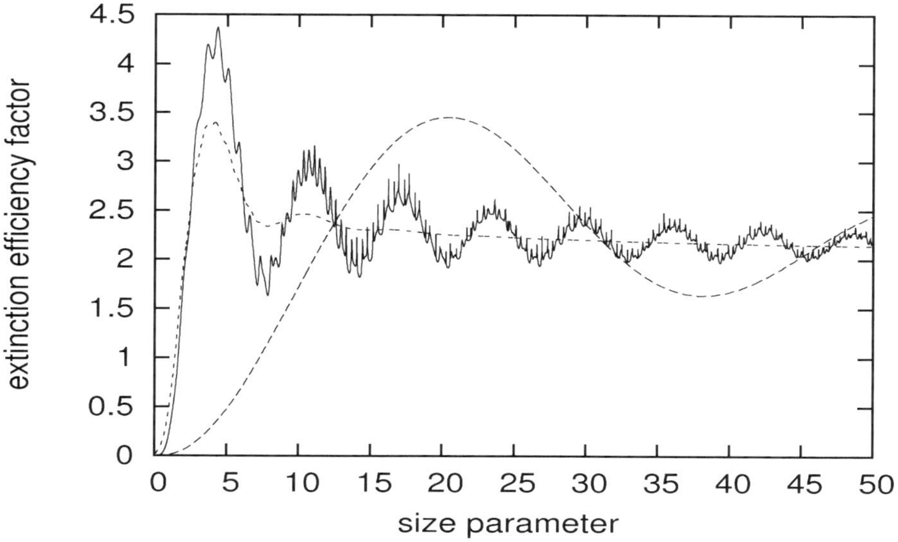

FIGURE 2.3: Plots of extinction efficiency factor against x. The relative refractive index of the sphere is (i) $m = 1.50$ (solid line), (ii) $m = 1.10$ (broken line), and (iii) $m = 1.5 + i0.1$ (dotted line).

More recently, simple empirical formulas for the extinction efficiency factor for nonabsorbing spheres in the $m - x$ domain $1 < m \leq 1.5$; $x \leq 50$ have been given by Bashkatova et al. (2001). The size parameter range has been divided in two parts.

$$(i)\ \ 0 < x \leq \frac{2.5869243m}{-0.99524473 + m}, \qquad (ii)\ \frac{2.5869243m}{-0.99524473 + m} \leq x \leq 50.$$

In the range (i), the extinction efficiency factor is parameterized as

$$Q_{ext} = \frac{A + Bx}{1 + Cx + Dx^2}, \tag{2.108}$$

where A, B, C and D are simple algebraic functions of m. In the range (ii), the extinction efficiency factor has been parameterized as

$$Q_{ext} = A^{'} e^{-B^{'} x} \sin(mxC^{'} + D^{'}) + 0.11m + 2.1. \tag{2.109}$$

Again, $A^{'}, B^{'}, C^{'}$ and $D^{'}$ are simple algebraic functions of m. Two sets of functions have been obtained. One for $1.0 \leq m \leq 1.25$ and the other for $1.0 \leq m \leq 1.30$.

Smart and Vand (1964) obtained the following empirical formula for scattering efficiency factor,

$$Q_{sca} = 2 + [Q_{sca}(1) - 2]\, f + R, \tag{2.110}$$

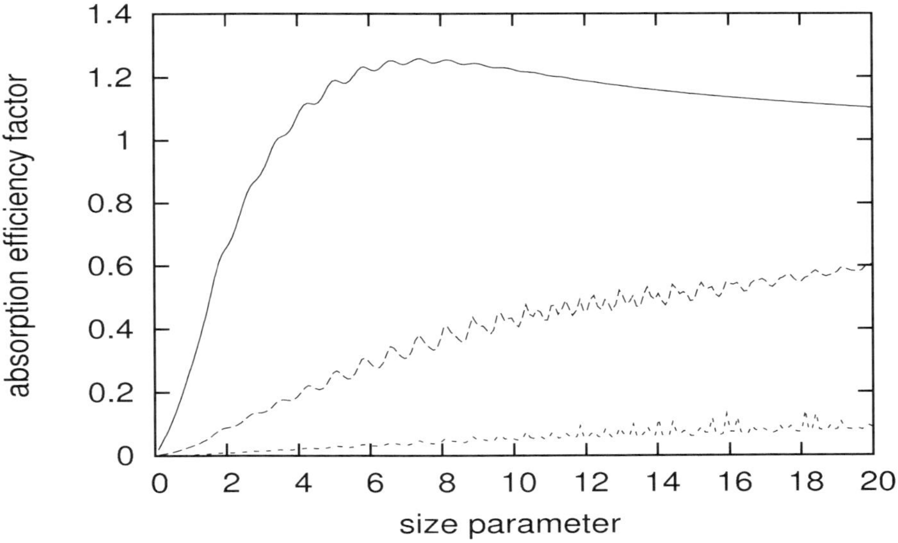

FIGURE 2.4: Plots of absorption efficiency factor Q_{abs} versus size parameter x. The relative refractive index is (i) $m = 1.5 + i0.10$ (solid line), (ii) $m = 1.5 + i0.01$ (large dashed line) and (iii) $m = 1.5 + i0.0010$ (small dashed line).

where

$$f = \frac{4m^2}{(m+1)^2},$$

and

$$R = \frac{2(f-1)}{z}\left(1-e^{-z}\right)\left[1 - ay^3e^{-y} - bu^2e^{-u^2}\right], \qquad (2.111)$$

with

$$z = 0.115f\rho, \quad y = \frac{2.38\rho}{f},$$

$$u = \frac{7.82(f-1)}{\rho}/2(m-1)f^4,$$

$$a = 0.3\left(1.75(f-1) + \sqrt{f-1}\right), \qquad (2.112)$$

and $b = 0.230f$. $Q_{sca}(1)$ is the value of Q_{sca} at the location of first maximum in the anomalous diffraction approximation (see, the Section 3.7). This gives errors to within 2% for $1.0 \leq m \leq 2.06$.

For a totally reflecting sphere, $m = \infty$. The following empirical form for the scattering efficiency factor,

$$Q_{sca} = Q_{ext} = 2 + 0.5x^{8/9} \qquad (2.113)$$

has been given by van de Hulst (1957) for such scatterers.

The influence of the absorption index on Q_{abs} has been examined by Hong (1977), Deirmendjian (1969) and Kattawar and Plass (1967), among others. The variation with size parameter for three values of the imaginary part in the refractive index is shown in Figure 2.4. The real part of the refractive index in all the three graphs is $n_r = 1.5$ and $n_i = 0.1,\ 0.01$ and, 0.001. Like Q_{ext}, the ripple structure is present in Q_{abs} as well. But the prominence of the ripple structure decreases as n_i increases.

The variation of Q_{abs} with the imaginary part of the refractive index is depicted in Figure 2.5. It can be seen that Q_{abs} increases rapidly with the increase in the absorption index and then falls to zero. This is because the high absorption index implies high conductivity. As a result, the electromagnetic wave is totally reflected and there is no absorption. The behaviour appears to have very little dependence on the real part of the refractive index.

One also defines the term, transport efficiency factor of scattering, as

$$Q^{tr}_{sca} = Q_{sca}(1-g). \qquad (2.114)$$

The resulting curves do not have numerous main oscillations because the oscillations of Q_{ext} and g are in phase. The main maximum is the same as that in the Q_{ext}. Any increase in the imaginary part of the refractive index results in a decrease of the transport scattering efficiency factor.

The asymmetry parameter (g) is another quantity from which the scatterer information can be decoded. Figure 2.6 exhibits variation of g against the size parameter x of a sphere. Plots for three relative refractive index values, $m = 1.10$, $m = 1.50$ and $m = 1.50 + i0.10$, are shown. Understandably, the asymmetry in the scattering pattern increases as x increases. The asymmetry parameter approaches its maximum value 1 as $x \to \infty$. For a real refractive index, this increase in g is oscillatory. However, as the imaginary part of the refractive index increases, the oscillations in the curve die down.

For a homogeneous spheres in the size parameter range $x \leq 50$, Bashkatova et al. (2001) give following simple algebraic expression for g. The x domain $0 \leq x \leq 50$, was divided in two regions. For the x regions

$$0 < x \leq 58.48979 - 49.978127 \exp\left(-3.7264027 m^{-14.934118}\right),$$

the asymmetry parameter is approximated as

$$g = \frac{\tilde{A} + \tilde{B}x}{1 + \tilde{C}x + \tilde{D}x^2}, \tag{2.115}$$

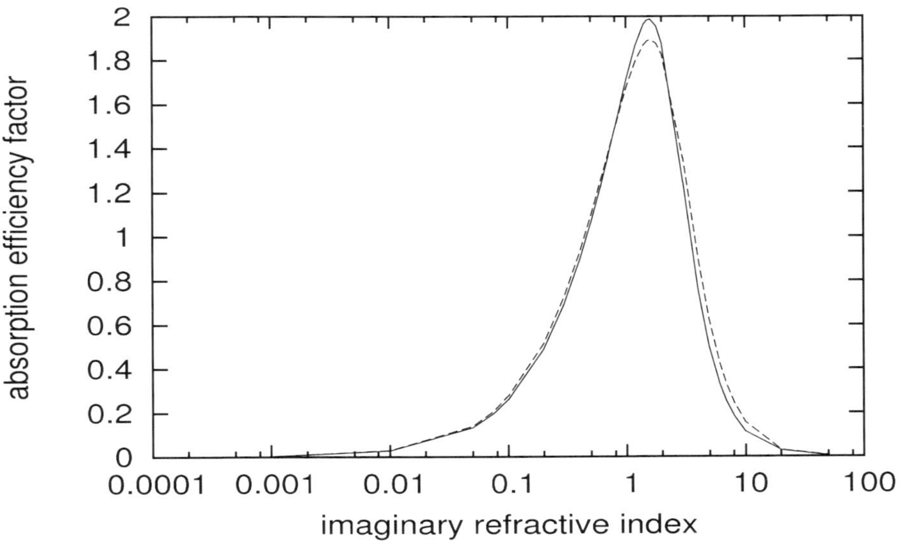

FIGURE 2.5: Plots of absorption efficiency factor Q_{abs} versus imaginary part of the refractive index (n_i). Solid line: $n_r = 1.1$, $x = 1.0$; dashed line: $n_r = 1.5$, $x = 1.0$.

where $\tilde{A}, \tilde{B}, \tilde{C}$ and $\tilde{D}$ are simple algebraic functions of m. For the x domain,

$$(58.48979 - 49.978127 \exp\left(-3.7264027 m^{-14.934118}\right) \leq x \leq 50.0,$$

the asymmetry parameter is,

$$g = \tilde{A}' + \tilde{B}' x + \frac{\tilde{C}'}{x^2}, \tag{2.116}$$

where $\tilde{A}'$, $\tilde{B}'$ and $\tilde{C}'$ are also simple algebraic functions of m. Numerical comparisons show that while the formula for Q_{ext} yields good agreement with Mie results, the accuracy of the formula for g can be termed only as qualitative.

2.2.5 Magnetic spheres

It can be seen from (2.24) and (2.25) that $a_n = b_n$, if $m\mu = \mu_1$. The meaning of the requirement $m\mu = \mu_1$ is that the relative magnetic permeability ($\mu_{rel} = \mu_1/\mu$) of a lossless sphere be equal to the relative refractive index m. A consequence of the fact that $a_n = b_n$ is that the backscattering efficiency Q_{back} , defined by (2.66), vanishes. It also follows that $S_1 = S_2$. Consequently, for linearly polarized incident radiation, the polarization of the scattered radiation will be the same for all scattering angles.

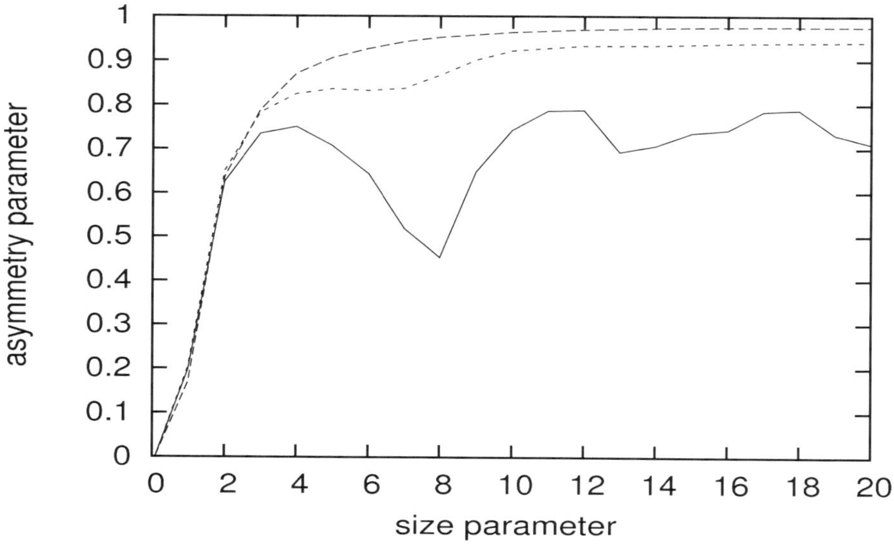

FIGURE 2.6: Plots of asymmetry parameter g versus size parameter x. The relative refractive index of the sphere is (i) $m = 1.5$ (solid line), (ii) $m = 1.1$ (large dashed line) and (iii) $m = 1.5 + i0.10$ (dashed line).

For small particles, only the leading terms a_1 and b_1 dominate the sum. For such particles, it can be shown that $a_1 = -b_1$ if (see, for example, Kerker et al. 1983),

$$\varepsilon = \frac{4 - \mu_{rel}}{2\mu_{rel} + 1} \tag{2.117}$$

where ε is the relative dielectric constant. The fact that $a_1 = -b_1$ implies that forward scattering is zero for small particles. That this indeed is the case has been verified by Mehta et al. (2006). The case $\mu_{rel} = 1$ is the trivial case where there is no scattering.

As the size of the scatterer increases, the condition for zero backscattering $\varepsilon = \mu_{rel}$ still remains valid, but the zero forward scattering condition no longer remains well founded, because higher-order terms in the series can no longer be ignored. Nevertheless, it is possible to find (ϵ, μ_{rel}) pairs which minimize the forward scattering. The following generalization of Kerker's condition has been arrived at (Garcia-Camara et al. 2010, 2011):

$$\varepsilon = \frac{\pi(4 - \mu_{rel}) - i\mathcal{V}k^3(\mu_{rel} - 1)}{\pi(2\mu_{rel} + 1) - i\mathcal{V}k^3(\mu_{rel} - 1)}, \tag{2.118}$$

where $\mathcal{V}$ is the volume of the particle.

2.2.6 Spheres in an absorbing host medium

If the host medium is absorbing, the absorption of the wave is not just due to the particle alone. The host medium also contributes. Initially the problem was addressed by Mundy et al. (1974), Chylek (1977), Bohren and Gilra (1979). Later, many others, including Quinten and Rotalski (1996), Sudiarta and Chylek (2001a, 2001b, 2002), Fu and Sun (2001), Yang et al. (2002), Videen and Sun (2003), Yin and Pilon (2006), Fu and Sun (2006), Frisvad et al. (2007) contributed to the solution.

The conventional Mie theory based on the far-field approximation can be generalized to the absorbing host medium by employing the following substitutions in the original Mie theory (see, for example, Yin and Pilon 2006). (i) Replace the real refractive index of the host medium N by the complex refractive index $N + iK$ and (ii) in efficiency factors (2.55) to (2.57), make the substitution,

$$\frac{2}{x^2} \rightarrow \frac{4K^2 \exp(-\eta r)}{(N^2 + K^2)[1 + (\eta a - 1)\exp(\eta a)]}, \tag{2.119}$$

where

$$\eta = 4\pi K/\lambda_0, \tag{2.119a}$$

and r is the distance of the detector from the centre of the sphere. Clearly, the efficiency factors defined in this way do not represent true radiative properties of the particle. They include effects of absorption by the host medium as well.

This complication can be circumvented by setting $r = a$ in (2.119) (Mundy et al. 1974). Thus, one can write

$$Q_{sca} = Q^0_{sca} e^{-\eta(r-a)}, \tag{2.120}$$

where

$$Q^0_{sca} = \frac{4K^2 \exp(-\eta a)}{(N^2 + K^2)[1 + (\eta a - 1)\exp(\eta a)]} \sum_{n=1}^{\infty} (2n+1)(|a_n|^2 + |b_n|^2), \tag{2.121}$$

is unattenuated scattering efficiency. For $Kx \ll 1$ and $K \ll N$, the coefficient is the same as in the conventional Mie theory. The efficiency factors based on far-field approximation are called apparent efficiency factors.

An alternative approach employs near-field approximation (Lebedev 1999, Sudiarta and Chylek 2001a, 2001b; Fu and Sun 2001) . The idea is to derive expressions for the efficiency factors based on the definitions (2.48). The relevant Poynting vectors are calculated at the surface of the sphere. The efficiency factors defined in this way are called the inherent efficiency factors.

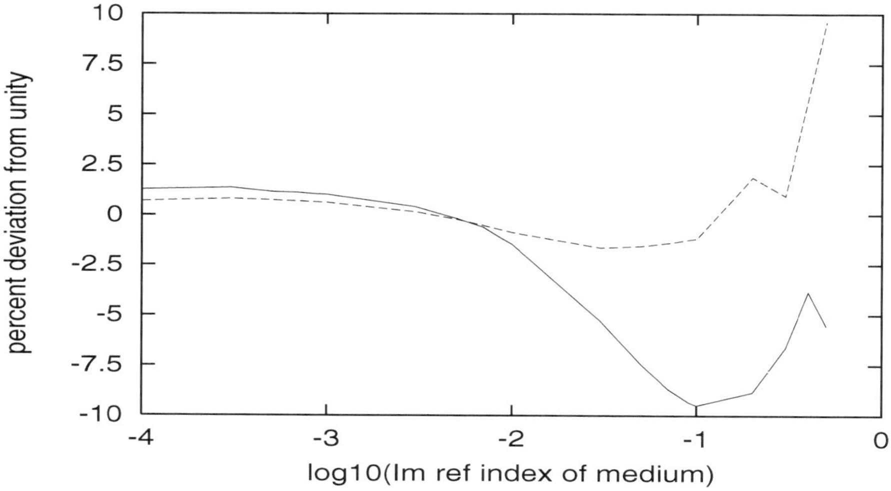

FIGURE 2.7: Departure of Q_{abs} from unity for a single spherical coal particle ($m = 1.7 + i0.04$) in an absorbing medium of refractive index ($m = 1.33 + im_{hi}$). The variation is against m_{hi}. The solid line is for $x = 15.708$ and the broken line is for $x = 62.832$. This figure was first published in Sharma and Jones (2003).

Using the Poynting vector, the expression for the W_{abs} and W_{sca} can be derived as,

$$W_{abs} = \frac{\pi|E_0|^2}{\omega\mu_1}\sum_{n=1}^{n=\infty}(2n+1)\mathrm{Im}A_n, \tag{2.122a}$$

and

$$W_{sca} = \frac{\pi|E_0|^2}{\omega\mu}\sum_{n=1}^{n=\infty}(2n+1)\mathrm{Im}B_n, \tag{2.122b}$$

where

$$A_n = \frac{|c_n|^2\psi_n(\beta)\psi_n^{'*}(\beta) - |d_n|^2\psi_n^{'}(\beta)\psi_n^*(\beta)}{k_1}, \tag{2.123a}$$

$$B_n = \frac{|a_n|^2\xi_n^{'}(\alpha)\xi_n^*(\alpha) - |b_n|^2\xi_n(\alpha)\xi_n^{'*}(\alpha)}{k}, \tag{2.123b}$$

with $\alpha = ka = 2\pi Na/\lambda_0$, $\beta = k_1 a = 2\pi am_1/\lambda_0$ and E_0 as the amplitude of the incident wave at the centre of the sphere. Thus, the rate at which the energy is incident is

$$I_i = \frac{2\pi a^2}{\eta^2}I_0\left[1+(\eta-1)e^{\eta}\right], \tag{2.124}$$

with

$$I_0 = \frac{N|E_0|^2}{2c\mu}. \tag{2.125}$$

Note that as $K \to 0$, $I_i = \pi a^2 I_0$. Numerical computations of this result for large-size particles show that the scattering efficiency factor approaches the Fresnel reflection coefficients for a plane surface at perpendicular incidence. That is,

$$\lim_{x\to\infty} Q_{sca} = \left|\frac{m-1}{m+1}\right|^2. \tag{2.126}$$

This may be contrasted with the asymptotic scattering efficiency factor in conventional Mie theory (see, for example, van de Hulst 1957):

$$\lim_{x\to\infty} Q_{sca} = 1 + \int_0^{\pi/2}\left(|r_1|^2 + |r_2|^2\right)\cos\theta\sin\theta d\theta, \tag{2.127}$$

where r_1, r_2 are the Fresnel reflection coefficients. It may be noted that the first term in (2.127), is the diffraction term. This term is absent in (2.126). The absence of this term in (2.126) results in an asymptotic extinction efficiency factor of 1 in contrast to the asymptotic extinction efficiency factor 2 in the conventional Mie theory.

Numerical comparisons of efficiency factors of spherical particles in an absorbing medium have been done for the far-field approximation and the near-field approximation by Yang et al. (2002), Randrianalisoa (2006), and Yin and Pilon (2006). The later two references include comparisons with conventional

Mie theory (real refractive index host medium) and with experimental data. The following observations have been made. (i) Deviations from conventional Mie theory can be significant, particularly for large particles and strong host absorption. (ii) Extinction efficiency factors obtained using far-field approximation and near-field approximation differ considerably. (iii) In calculations of the radiative transfer, the differences get masked.

An important observation made by Yin and Pilon (2006) is that the absorbing medium has little effect on Q_{abs}. A similar observation was made by Sharma and Jones (2003) in the context of scattering of light by highly absorbing coal particles embedded in a water drop. The imaginary part of the host refractive index, m_{hi}, was varied from 0.0001 to 0.5. The resulting Q_{abs} can be seen to be nearly constant at $Q_{abs} = 1$. The maximum deviation from $Q_{abs} = 1$ is about 10%. Two plots in the Figure 2.7 are for $x = 15.708$ and $x = 62.832$.

Figure 2.8 shows variation in the albedo and asymmetry parameters with the imaginary part of the host medium refractive index. It can be seen that their variation is very weak when the host medium absorption is weak. A sharp decrease in the asymmetry parameter is observed for large particles and stronger host medium absorption. This is expected because the reflection increases as the imaginary part of the refractive index increases.

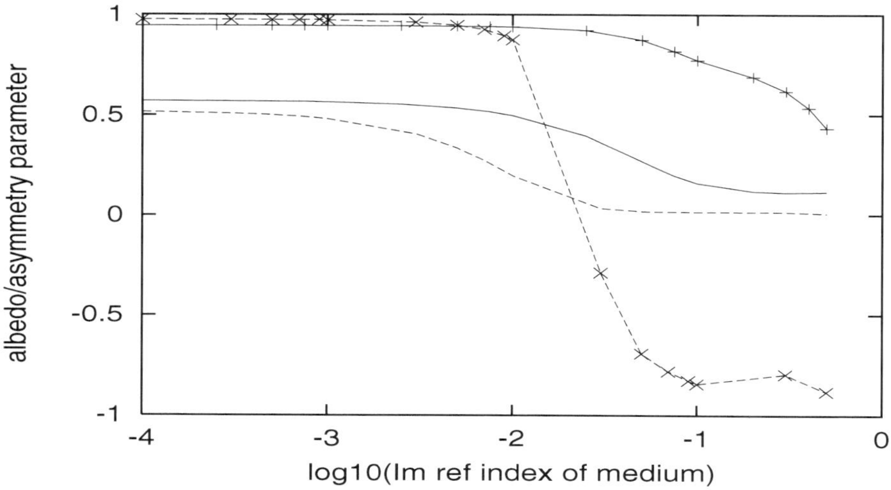

FIGURE 2.8: Single spherical coal particle in an absorbing medium of refractive index $m = 1.33 + im_{hi}$. Albedo: Solid line ($x = 15.708$), dashed line ($x = 62.832$). Asymmetry parameter: Solid line with pluses ($x = 15.708$) and the dashed line with crosses ($x = 62.832$). This figure was first published in Sharma and Jones (2003).

2.2.7 Charged spheres

Many particles of interest in nature are electrically charged. For example, atmospheric and interstellar grains are believed to include charged particles. The first attempt to solve this scattering problem appears to be attributed to Bohren and Hunt (1977). Scrutinizing the scattering of electromagnetic waves by a uniformly charged metallic sphere, it was revealed that the scattering coefficients of a charged particle are different from those of an uncharged particle. Klacka and Kocifaj (2007, 2010) revisited the solution and provided following expressions for the scattering coefficients a_n and b_n:

$$a_n = \frac{m\mu\psi_n(mx)\psi_n'(x) - \mu_1\psi_n(x)\psi_n'(mx) + i\omega k^{-1}\mu\mu_1\sigma_s\psi_n'(x)\psi_n'(mx)}{m\mu\psi_n(mx)\xi_n'(x) - \mu_1\xi_n(x)\psi_n'(mx) + i\omega k^{-1}\mu\mu_1\sigma_s\xi_n'(x)\psi_n'(mx)}, \tag{2.128a}$$

and

$$b_n = \frac{m\mu\psi_n(x)\psi_n'(mx) - \mu_1\psi_n'(x)\psi_n(mx) - i\omega k^{-1}\mu\mu_1\sigma_s\psi_n(x)\psi_n(mx)}{m\mu\psi_n'(mx)\xi_n(x) - \mu_1\xi_n'(x)\psi_n(mx) - i\omega k^{-1}\mu\mu_1\sigma_s\xi_n(x)\psi_n(mx)}, \tag{2.128b}$$

where σ_s is the surface conductivity of the sphere. For $\sigma_s = 0$, the Mie coefficients are identical to those for a uncharged sphere.

Computation of scattering phase function of a charged sphere shows it to be substantially different from the corresponding phase function of an uncharged sphere (Hu and Xie 2015). For each particle size, a threshold surface conductivity exists, beyond which the angular scattering depends only on surface conductivity. The differences maximize at backward angles. For materials having a small imaginary part in the refractive index, the attenuation of electromagnetic radiation was found to be about 10 times more efficient when set side by side with to neutral sphere.

A model in which charge distribution on the surface of the sphere is assumed to be distributed uniformly only over a cap-like region, defined by an angle θ_0 with the vertical axis, has been examined by Zhou et al. (2005). Thus, for $\theta_0 = \pi/2$, the charge is uniformly distributed over the upper half and $\theta_0 = \pi$ amounts to electric charge distribution over the whole surface. For particles satisfying the condition $x \ll 1$, the analytic expression for $i(\theta)$ is,

$$i(\theta) = \frac{k^6}{(4\pi\varepsilon_0)^2}\left[\left(4\pi\varepsilon_0 a^3\left|\frac{\varepsilon_1 - \varepsilon_0}{\varepsilon_1 + 2\varepsilon_0}\right]\cos\theta\cos\varphi + \frac{\sigma(\varepsilon_1 - \varepsilon_0)}{\varepsilon_0 E_0}\pi a^3 \sin\theta_0\right)^2 + \left(4\pi\varepsilon_0 a^3\left|\frac{\varepsilon_1 - \varepsilon_0}{\varepsilon_1 + 2\varepsilon_0}\right|\sin\varphi\right)^2\right]. \tag{2.129}$$

The efficiency factors are:

$$Q_{sca} = \frac{8}{3}k^4 a^4\left|\frac{\varepsilon_1 - \varepsilon_0}{\varepsilon_1 + 2\varepsilon_0}\right|^2 + \frac{1}{4}k^4 a^4 \frac{\sigma_s^2(\varepsilon_1 - \varepsilon_0)^2 \sin^2\theta_0}{\varepsilon_0^4 E_0^2}, \tag{2.130}$$

and

$$Q_{abs} = 12ak\varepsilon_{img}\left|\frac{\varepsilon_0}{\varepsilon_1+\varepsilon_0}\right|^2, \tag{2.131}$$

where ε_{img} is the imaginary part of the relative permittivity. As usual ε_0 and ε_1 are the permittivities of the vacuum and the particle.

Based on the Rayleigh approximation, Li et al. (2012, 2014) have studied the attenuation of electromagnetic waves by small charged particles. Analytic formulas have been derived and it is shown that the impact of charges on anisotropic particles is much greater than that on the isotropic particles.

Electromagnetic wave scattering by a polydispersion of charged particles has been studied by Kocifaj et al. (2011). The total optical thickness and scattering phase function obtained for charged particles were contrasted with those obtained for electrically neutral particles. It was concluded that the surface charges on cosmic dust particles could modify both of the quantities significantly. Kocifaj and Klacka (2012) have also studied near-field intensity distribution and showed that the scattering from a charged sphere differs from the scattering by an uncharged sphere appreciably only if $x > 1$.

2.2.8 Chiral spheres

The first attempt at studying scattering of electromagnetic waves by a chiral sphere seems to be attributable to Gordon (1972). Exact analytic solutions for the problem were derived by Bohren (1974). A plane wave entering a chiral medium splits into two parts. The left-handed circularly polarized wave travels at a phase velocity which is different from the phase velocity of the right handed-circularly polarized wave causing the rotation of the plane of polarization. The scattering functions $S_1(\theta)$ and $S_2(\theta)$ are formally the same. The coefficients a_n and b_n are (Bohren and Huffman 1983):

$$a_n = \frac{D_n(m_R,x)A_n(m_L,x)+D_n(m_L,x)A_n(m_R,x)}{C_n(m_L,x)V_n(m_R,x)+D_n(m_L,x)C_n(m_R,x)}, \tag{2.132a}$$

and

$$b_n = \frac{C_n(m_L,x)B_n(m_R,x)+C_n(m_R,x)B_n(m_L,x)}{C_n(m_L,x)D_n(m_R,x)+D_n(m_L,x)C_n(m_R,x)}. \tag{2.132b}$$

The scattering functions S_3 and S_4, which were zero for a homogeneous sphere, now become:

$$S_3(\theta) = -S_4(\theta) = \sum_{n=1}^{\infty}\frac{2n+1}{n(n+1)}c_n(\pi_n+\tau_n), \tag{2.133}$$

where

$$c_n = -d_n = \frac{C_n(m_R,x)A_n(m_L,x)-C_n(m_L,x)A_n(m_R,x)}{C_n(m_L,x)D_n(m_R,x)+D_n(m_L,x)C_n(m_R,x)}, \tag{2.134}$$

with

$$m_L = \frac{N_L}{N} \text{ and } m_R = \frac{N_R}{N}. \tag{2.135}$$

The efficiency factors for the left circularly polarized wave are,

$$Q_{sca,L} = \frac{2}{x^2} \sum_{n=1}^{\infty} (2n+1) \left[|a_n|^2 + |b_n|^2 + 2|c_n|^2 - 2\text{Im}\left((a_n + b_n)c_n^*\right) \right], \tag{2.136}$$

$$Q_{ext,L} = \frac{2}{x^2} \sum_{n=1}^{\infty} (2n+1) \text{Re}\left[a_n + b_n - 2ic_n\right]. \tag{2.137}$$

The corresponding efficiency factors for a right circularly polarized wave can be obtained by changing the sign of the last term on the right-hand side in the above equations. The quantities A_n, B_n C_n and D_n are given by (2.28) and (2.29).

Scattering by a chiral sphere in chiral medium has been examined by Hinders and Rhodes (1992), by chiral spheroids by Cooray and Ciric (1993) and by Lakhtakia et al. (1985), by a stratified chiral sphere by Jaggard and Liu (1999) and Li et al. (1999). Analytic expressions for the scattering coefficients for a beam of arbitrary shape have been obtained by Wu et al. (2012).

2.2.9 Layered spheres

The problem of scattering of electromagnetic waves by a single layer coated sphere was first solved by Aden and Kerker (1951). Independently, Güttler (1952) and Shifrin (1952) also obtained the solution to this problem. A sphere with more than one layer was addressed in a similar way by Mikulski and Murphy (1963), Wait (1963), Ferdinandov (1967), Pontikis (1968), Tuomi (1980), and Lopatin and Sid'ko (1988). The solution to the problem of an arbitrary number of layers can be found in Kerker (1969), Bhandari (1985), and Shore (2015). Kai and Massoli (1994) developed a computer program which could accommodate up to 10,000 shells. The derivation of expressions for characteristics of concentric spheres was extended to non-concentric spheres by Ngo et al. (1996). The solution for a radially inhomogeneous sphere was obtained by Perelman (1996).

A coated sphere serves as a model for diverse scattering objects—in nature as well as in the laboratory. Such particle models are needed in meteorology, atmospheric optics, physical chemistry, astrophysics and biology. Astronomers found that the simple core-mantel models for interstellar dust particles were not good enough to explain the interstellar extinction and polarization. This has led to consideration of multilayered spheres as the models for dust grains (Voshchinnikov and Mathis 1999). Three layered spheres have been used to model biological cells with a thin surface layer mimicking the cell shell, core (nucleus) and cytoplasm as intermediate layer.

The formulas for angular scattering and efficiency factors by any spherically symmetric particle have the same form as that for a homogeneous sphere. The only distinction is that the scattering coefficients are different. For an L layer model with the core as the first layer, the distance $r = a_1$ defines the interface between the core and the first coating (layer 2). Similarly, the interface at $r = a_j$ is the interface between layers j and $j+1$. The interface at $r = a_L$ defines the interface between layer L and the host medium. The refractive index of j-th layer is defined as N_j, and N_{L+1} is the refractive index of the host medium. Following Shore (2015), the scattering coefficient a_n may be written as:

$$a_n = \frac{\mu_{L+1} N_L \psi_n^{'}(x_L) - \mu_L N_{L+1} \psi_n(x_L) R_{S,n}^L}{\mu_{L+1} N_L \xi_n^{'}(x_L) - \mu_L N_{L+1} \xi_n(x_L) R_{S,n}^L}, \tag{2.138}$$

where

$$y_j = k_j a_j, \ x_j = k_{j+1} a_j, \ \frac{y_j}{x_j} = \frac{k_j}{k_{j+1}} = \frac{N_j}{N_{j+1}}, \tag{2.138a}$$

and

$$R_{S,n}^j = \frac{\psi_n^{'}(y_j) + \chi_n^{'}(y_j) S_n^j}{\psi_n(x_j) + \chi_n(x_j) S_n^j}, \tag{2.138b}$$

with $S_n^1 = 0$, and

$$S_n^{j+1} = \frac{\mu_{j+1} N_j \psi_n^{'}(x_j) - \mu_j N_{j+1} \psi_n^{'}(x_j) R_{S,n}^j}{\mu_{j+1} N_j \chi_n(x_j) - \mu_j N_{j+1} \chi_n(x_j) R_{S,n}^j}, \quad j = 1, 2, ...L-1, \tag{2.138c}$$

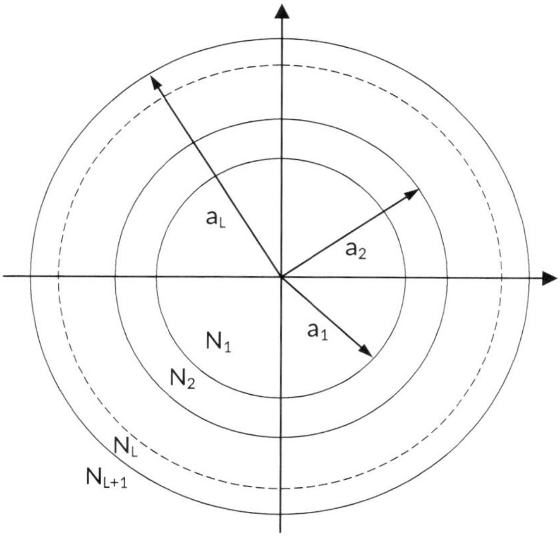

FIGURE 2.9: Geometry of a layered sphere.

The expression for the scattering coefficient b_n is:

$$b_n = \frac{\mu_L N_{L+1}\psi_n^{'}(x_L) - \mu_{L+1} N_L \psi_n(x_L) R_{T,n}^L}{\mu_L N_{L+1}\xi_n^{'}(x_L) - \mu_{L+1} N_L \xi_n(x_L) R_{T,n}^L}, \tag{2.139}$$

where

$$R_{S,n}^j = \frac{\psi_n^{'}(y_j) + \chi_n^{'}(y_j) T_n^j}{\psi_n(y_j) + \chi_n(y_j) T_n^j}, \tag{2.139a}$$

with $T_n^1 = 0$, and

$$T_n^{j+1} = \frac{\mu_j N_{j+1}\psi_n^{'}(x_j) - \mu_{j+1} N_j \psi_n^{'}(x_j) R_{T,n}^j}{\mu_j N_{j+1}\chi_n(x_j) - \mu_{j+1} N_j \chi_n(x_j) R_{T,n}^j}, \qquad j = 1, 2, ...L-1, \tag{2.139b}$$

Relevant formulas for a coated sphere can be easily deduced from the above expressions. Let the magnetic permeabilities in all the media be unity. Then, defining a_1, m_1 as the radius and relative refractive index of the core, a_2, m_2 as the radius and relative refractive index of the coating, the Mie coefficient a_n becomes,

$$a_n = \frac{m_2\psi_n^{'}(y) - \psi_n(y) R_{s,n}^2}{m_2\xi_n^{'}(y) - \xi_n(y) R_{s,n}^2}, \tag{2.140}$$

where $x = ka_1$, $y = ka_2$,

$$R_{s,n}^2 = \frac{\psi_n^{'}(m_2 y) - \chi_n^{'}(m_2 y) S_n^2}{\psi_n(m_2 y) - \chi_n(y) S_n^2}, \tag{2.141}$$

with

$$S_n^2 = \frac{m_1\psi_n^{'}(m_2 x) - m_2\psi_n(m_2 x) R_{s,n}^1}{m_1\chi_n^{'}(m_2 x) - m_2\chi_n(m_2 x) R_{s,n}^1}, \tag{2.142}$$

and

$$R_{s,n}^1 = \frac{\psi_n^{'}(m_1 x)}{\psi_n(m_1 x)}. \tag{2.143}$$

The scattering coefficient a_n can then be expressed as,

$$a_n = \frac{\psi_n(y)[\psi_n^{'}(m_2 y) - A_n\chi_n^{'}(m_2 y)] - m_2\psi_n^{'}(y)[\psi_n(m_2 y) - A_n\chi_n(m_2 y)}{\xi_n(y)[\psi_n^{'}(m_2 y) - A_n\chi_n^{'}(m_2 y)] - m_2\xi_n^{'}(y)[\psi_n(m_2 y) - A_n\chi_n(m_2 y)}, \tag{2.144}$$

with

$$A_n = \frac{m_2\psi_n(m_2 x)\psi_n^{'}(m_1 x) - m_1\psi_n^{'}(m_2 x)\psi_n(m_1 x)}{m_2\chi_n(m_2 x)\psi_n^{'}(m_1 x) - m_1\chi_n^{'}(m_2 x)\psi_n(m_1 x)}. \tag{2.145}$$

In the same way, it can be shown that

$$b_n = m_2\frac{\psi_n(y)[\psi_n^{'}(m_2 y) - B_n\chi_n^{'}(m_2 y)] - m_2\psi_n^{'}(y)[\psi_n(m_2 y) - B_n\chi_n(m_2 y)}{\xi_n(y)[\psi_n^{'}(m_2 y) - B_n\chi_n^{'}(m_2 y)] - m_2\xi_n^{'}(y)[\psi_n(m_2 y) - B_n\chi_n(m_2 y)}, \tag{2.146}$$

with

$$B_n = \frac{m_2\psi_n(m_1x)\psi_n'(m_2x) - m_1\psi_n'(m_1x)\psi_n(m_2x)}{m_2\chi_n'(m_2x)\psi_n(m_1x) - m_1\chi_n(m_2x)\psi_n'(m_1x)}. \tag{2.147}$$

A hollow sphere or a bubble is a special case where $m_1 = N$. Note that for $m_1 = m_2$, $A_n = B_n = 0$. Expressions for a_n and b_n reduce to those for a homogeneous sphere.

Expressions for the efficiency factors are:

$$Q_{ext} = \frac{2}{x_L^2}\sum_{n=1}^{\infty}\Big[(2n+1)\mathrm{Re}(a_n + b_n)\Big], \tag{2.148a}$$

$$Q_{sca} = \frac{2}{x_L^2}\sum_{n=1}^{\infty}\Big[(2n+1)(|a_n|^2 + |b_n|^2)\Big], \tag{2.148b}$$

and

$$Q_{abs} = Q_{ext} - Q_{sca} = \frac{2}{x_L^2}\sum_{n=1}^{\infty}(2n+1)\left[\frac{1}{2} - \left|a_n - \frac{1}{2}\right|^2 - \left|b_n - \frac{1}{2}\right|^2\right]. \tag{2.148c}$$

But for a difference in the factor outside the summation sign, these expressions are formally identical to those for a homogeneous sphere. The factor which was $2/(k^2a^2)$ for a homogeneous sphere outside the summation sign now changes to $2/(k^2a_L^2)$, where a_L is the radius of the last coating. The same replacement is needed in the expression for the sphere asymmetry parameter to obtain the corresponding expression for a layered sphere.

The dependence of the efficiency factors on the number of layers, their refractive index and thickness has been examined by many researchers in varied contexts (Kerker 1969, Brunsting and Mullaney 1971, Volkov and Kovach 1990, Kitchen and Zaneveld 1992, Sharma and Somerford 1992, Quirantes and Bernard 2004, Small et al. 2005). A detailed survey of studies on two layered spheres can be found in Babenko et al. (2003). If the core is absorbing and the shell is nonabsorbing, then the absorption cross section oscillates about an average value, σ_{abs}^{asymp}, as the shell thickness increases. The amplitude of oscillation is dependent on the core radius and optical constants (Fenn and Oser 1965). The smaller the core size, the larger the shell thickness needs to be to reach the asymptotic value of the cross section.

Some empirical formulas obtained for the cross section are:

(i) For $1.1 \leq n_{1r} \leq 1.7$, $x_1 < 15$, $n_{2r} = 1.33$, $\lambda = 0.5\ \mu m$ and $n_{2i} = 0$, the following empirical formula was obtained by Prishivalko et al. (1975):

$$\sigma_{abs}^{asymp} = \alpha n_{1i}^2 + \beta n_{1i} + \gamma n_{1i}^{\delta} n_{1r} + \epsilon. \tag{2.149}$$

The values of the coefficients are a_1 and n_{1i} dependent. The accuracy is to within 5% when $a_1 = 0.1\ \mu m$ and $0 \leq n_{1i} \leq 0.2$ ($\alpha = 0, \beta = 0.0485, \gamma = 0.0639, \delta = 0.0639, \epsilon = -0.56 \times 10^{-4}$). The error seems to increase with increasing a_1. For $a_1 = 1.0\ \mu m$ and $0 \leq n_{1i} - 1 \leq 0.05$, the error is less than 20%.

(ii) Dependence of the asymptotic absorption cross section on core size may be cast as,

$$\sigma_{abs}^{asymp} = A(n_{1r}, n_{1i}) x_1^{B(n_{1r}, n_{1i})}. \tag{2.150}$$

The values of the coefficients A and B can be found in Babenko et al. (2003) as well. The errors in (2.150) are to within 1.5% for $0.1 \leq a_1 \leq 1.0\ \mu m$ and $n_{1r} = 2$. For larger core spheres, the results agree only qualitatively. In a reverse situation where $m_2 > m_1$, a decrease in absorption efficiency is observed.

Small et al. (2005) considered particles in the size range $25\ nm \leq a \leq 0.5\ \mu m$, and core and coatings with refractive indices between 1.35 to 1.65, respectively, corresponding to particles used in commercial plastics. The host refractive index was taken to be 1.5 and the incident radiation wavelength was $550 nm$. It was concluded that (i) a coating to a polymer particle leads to a significant reduction in its scattering cross section if the coating is such that it reduces volume averaged dielectric contrast between the particle and the host medium. (ii) The scattering cross section and angular scattering are not very sensitive to morphological details of the particle if the volume averaged dielectric constant of the particle is not close to that of the surrounding medium. (iii) When the obstacle is large, homogeneous particles provide the least scattering.

Some conclusions by Volkov and Kovach (1990) in the context of aerosols are: (i) if there is a highly absorbent external layer, the morphology of internal layers is not of much significance. (ii) If the product of the outer layer thickness and imaginary part of the refractive index exceeds a threshold value, the sphere can be regarded as homogeneous consisting of outer layer material. (iii) If a strongly absorbing layer lies within the sphere, the scattering behaviour is determined by this layer and subsequent layers.

Sharma and Somerford (1990) depicted graphs for Q_{ext} vs. $q_r = a_1/a_2$. Thus, $q_r = 1$ corresponds to a homogeneous sphere of core material and $q_r = 0$ corresponds to a homogeneous sphere of coating material. The relative refractive indices of core and coating with respect to host medium were taken to be $m_1 = 2.0$ and $m_2 = 1.02$. These values are typical of particles found in cohesive sediments with a coating of biological material. For small particle sizes, variation of Q_{ext} with q_r seems to agree broadly with conclusion (i) of Small et al. (2005).

2.2.10 Debye series

The Debye series expansion is a parallel approach to the Mie series expansion. It decomposes the partial wave amplitudes in terms of diffraction, reflection and transmission components and provides a direct understanding of the physics involved. The series was proposed by Debye (1908) in the context of light scattering by a circular cylinder. It was extended to a homogeneous sphere by van der Pol and Bremmer (1937a, 1937b), to a coated sphere by Lock et al. (1994), and to a multilayered sphere by Lock (2005), to a tilted cylinder by Lock and Adler (1997). The Debye series for a shaped beam incidence on a sphere was given by Gouesbet (2003) and for spheroids (plane wave as well as shaped beam incidence) by Xu et al. (2010a). The Debye series for irregularly shaped particles was also developed by Xu et al. (2010b). The Debye series for interior amplitudes for a sphere is by Lock (1988).

Although the Debye series gives a clear physical picture of the scattering process, the convergence of the series is comparatively slow. Nussenzweig (1969a, 1969b) developed a method known as the complex angular momentum (CAM) method to improve the convergence. But even so, the numerical computation of the Debye series is more difficult in comparison to the Mie series.

2.3 Resonances of the Mie coefficients

Resonances in the Mie coefficients occur when the denominators in a_n or b_n vanish. The resonance frequencies are governed by size, shape and refractive index of the sphere. Hence, these resonances are known as morphology dependent resonances (MDRs). To acquire a better understanding of MDRs, let us recast a_n (given in (2.24)) in the following form:

$$a_n = \frac{p_n}{p_n + iq_n} = \frac{p_n^2}{p_n^2 + q_n^2} - i\frac{p_n q_n}{p_n^2 + q_n^2}, \tag{2.151}$$

where

$$p_n = m\psi_n(mx)\psi_n^{'}(x) - \psi_n(x)\psi_n^{'}(mx), \tag{2.152a}$$

and

$$q_n = -m\psi_n(mx)v_n^{'}(x) + v_n(x)\psi_n^{'}(mx), \tag{2.152b}$$

with $v_n(x) = -xy_n(x)$. A similar decomposition for b_n, given by (2.25), gives

$$b_n = \frac{r_n}{r_n + is_n} = \frac{r_n^2}{r_n^2 + s_n^2} - i\frac{r_n s_n}{r_n^2 + s_n^2}, \tag{2.153}$$

with

$$r_n = m\psi_n(x)\psi_n^{'}(mx) - \psi_n(mx)\psi_n^{'}(x), \tag{2.154a}$$

and

$$s_n = -mv_n(x)\psi_n^{'}(mx) + \psi_n(mx)v_n^{'}(x). \qquad (2.154b)$$

If the refractive index is real, the functions p_n, q_n, r_n, and s_n are all real. As a consequence, the denominators of a_n or b_n cannot vanish completely. The vanishing of the denominator is possible only if the refractive index of the particle has an imaginary part in it. For a particle with real refractive index, a resonance will occur when either q_n or s_n vanishes. When this happens, the real parts of a_n or b_n reach their maximum possible value of unity and the imaginary parts vanish. For each a_n and b_n, there is a sequence of x values for which resonances may occur. The identification of a resonance is done by labelling it with a type of mode, a or b, with mode number n as subscript and a superscript is used to identify the sequential order of x.

Several authors have derived analytic expressions for various features of the resonances. While Probert-Jones (1984) accomplished expressions for locations and width of resonances for large dielectric spheres, Lam et al. (1992) obtained explicit formulas for the position, width and strength of resonances. Exact analytic formulas for predicting the width of these resonances have also been obtained by Johnson (1993) in an analogy with quantum mechanical shape resonance theory. Chylek (1976, 1990) and Chylek et al. (1978) have also investigated resonances in detail. For sufficiently small x, the spacing between two resonances a_n^1 and a_{n+1}^1 can be shown to be

$$\Delta x = \frac{x}{n}\frac{\tan^{-1}\left(\sqrt{(mx/n)^2 - 1}\right)}{\sqrt{(mx/n)^2 - 1}} \quad \text{if } |x - n| \gg 1/2, \qquad (2.155a)$$

and

$$\Delta x = \frac{\tan^{-1}\left(\sqrt{m^2 - 1}\right)}{\sqrt{m^2 - 1}} \quad \text{if } \frac{x}{n} \sim 1. \qquad (2.155b)$$

It is, therefore, possible to find the size parameter by an appropriate numerical program. A comparison of Δx computed from equations (2.155a) and (2.155b) with those calculated using Mie theory show that the accuracy of the derived equation is about 1%, as long as $|x - n| \geq 4$.. A computer program for calculation of resonance location was developed by Cantrell (1988).

2.4 Other shapes

An infinite cylinder is another shape which has been studied very widely. Its closeness to infinite length is specified in terms of the aspect ratio defined as length/diameter. This should be large. A value greater than 10 is generally taken to be a good approximation.

The exact solution to the problem was first obtained by Rayleigh (1881) for perpendicular incidence. Wait (1955) solved the problem for oblique incidence. Blank (1955) and Burberg (1956) also solved the problem around the same time. Kerker and Matijevic (1961) accomplished the solution for two concentric cylinders at perpendicular incidence and Shah (1970) extended the solution to oblique incidence. The problem of scattering by a multilayered cylinder was solved by Yeh and Lindgreen (1977) for normal incidence and by Barabas (1987) for a tilted cylinder. Kai and D'Alessio (1995) solved the problem of a radially inhomogeneous cylinder by considering it as a finely stratified cylinder. The program developed by them could accommodate up to 80,000 layers.

Following empirical parametrization for Q_{ext}, g was obtained by Bashkotova et al. (2001) for nonabsorbing cylinders at perpendicular incidence. The x range, $x \leq 50$, has been divided in two regions. For

$$0 < x \leq [-0.0074728439 + 0.56203885 \log(m)]^{-1},$$

extinction efficiency has the form

$$Q_{ext}(x, m) = A + B \cos\left(C + Dx\right), \tag{2.156}$$

where A, B, C and D are simple algebraic and trigonometric functions of m. For the x range,

$$[-0.0074728439 + 0.56203885 \log(m)]^{-1} \leq x \leq 50.0,$$

the extinction efficiency factor is of the form

$$Q_{ext}(x, m) = A \exp(-Bx) \sin(C + D\ mx) + E, \tag{2.157}$$

where A, B, C, D and E are also simple algebraic functions of m.

The asymmetry parameter in the x range,

$$0 < x \leq \left[0.50696001 - 0.92345444m + 0.45971355m^2\right]^{-1},$$

has been parametrized as

$$g(x, m) = \frac{A + Bx}{1.0 + Cx + Dx^2}, \tag{2.158}$$

TABLE 2.1: Scatterer description and references

Scattering description	Reference
Dielectric-disk arrays	Schaudt et al. 1991
Uniaxial anisotropic sphere	Wong and Chen 1992
Sphere with an irregular inclusion	Videen et al. 1995
Concentric optically active sphere	Kim and Chang 2004
Uniaxial left-handed material	Geng et al. 2006
Pits and holes in a metal layer	Brok 2007
Chebyshev particles	Petrov et al. 2007
Uniaxial anisotropic coating	Geng et al. 2009
Nearly spherical nanoparticles	Xie et al. 2010
Large-sized chiral cylinder	Shang et al. 2014
Monolayer of nematic droplets	Loiko et al. 2016

and in the size parameter range,

$$\left[0.50696001 - 0.92345444m + 0.45971355m^2\right]^{-1} \leq x \leq 50.0,$$

it has been parametrized as

$$g(x, m) = A + Bx + \frac{C}{x^2}, \tag{2.159}$$

where A, B, C and D are again simple algebraic functions of m. The expressions for A, B, C in (2.156) to (2.159) are not identical. The range of applicability of these formulas is $1.0 \leq m \leq 1.5$ and $0 < x \leq 50.0$. Numerical comparisons of these formulas show the extinction efficiency factor to be in very good agreement with the exact results. The accuracy of the asymmetry parameter formula is lower in comparison to Q_{ext}. It yields better approximation for soft cylinders.

2.5 Integral equation method

Let us begin with equation (1.32), which can be rewritten as,

$$\left(\nabla^2 + k^2\right)\mathbf{E}(\mathbf{r}) = -(m^2(\mathbf{r}) - 1)k^2\mathbf{E}(\mathbf{r}) - \nabla\mathbf{E}(\mathbf{r}).\nabla \log m^2(\mathbf{r})). \quad (2.160)$$

The vector integral equation corresponding to (2.160) can be expressed as (Saxon 1955),

$$\mathbf{E}(\mathbf{r}) = \mathbf{E}_{inc}(\mathbf{r}) + k^2 \int_{v'} \mathbf{E}(\mathbf{r}')(m^2(\mathbf{r}') - 1)G(\mathbf{r}, \mathbf{r}')d\mathbf{r}'$$

$$+ \int_{v'} \nabla'[\nabla'.\mathbf{E}(\mathbf{r}')(m^2(\mathbf{r}') - 1)]G(\mathbf{r}, \mathbf{r}')d\mathbf{r}', \quad (2.161)$$

where v' is the particle volume. The total field at anywhere in space is

$$\mathbf{E} = \mathbf{E}_{inc} + \mathbf{E}_{sca}. \quad (2.162)$$

The scattered field in the far-field region can be written as (Saxon 1955, Bayvel and Jones 1981, Klett and Sutherland 1992),

$$\mathbf{E}_{sca} = E_0 \mathbf{f} \frac{e^{ikr}}{r}, \quad (2.163)$$

where

$$\mathbf{f} = k^2[1 - \hat{r}(\hat{r}.)]\mathbf{I}, \quad (2.164)$$

is the scattering amplitude, with

$$\mathbf{I} = \frac{1}{4\pi} \int_{v'} \frac{\mathbf{E}(\mathbf{r}')}{E_0} \left(m^2(\mathbf{r}') - 1\right) e^{-ik_s.r'} d^3r'. \quad (2.165)$$

The internal electric field needed in (2.165) is not known a priori. This has resulted in numerous approximations based on judicious guesses for $\mathbf{E}(\mathbf{r})$. If $\mathbf{E}(\mathbf{r})$ is approximated as $\hat{\mathbf{e}}_i E_0 e^{i\mathbf{k}_i.\mathbf{r}}$, the scattering amplitude is Rayleigh–Gans scattering amplitude, and can be expressed as

$$\mathbf{f} = \frac{k^2}{4\pi}[1 - \hat{r}(\hat{r}.)]\hat{\mathbf{e}}_i \int_{v'} (m^2(\mathbf{r}') - 1)e^{i\mathbf{q}.\mathbf{r}'} d^3r', \quad (2.166)$$

where $\mathbf{q} = \mathbf{k}_i - \mathbf{k}_s$. The quantity $[1 - \hat{r}(\hat{r}.)]\hat{\mathbf{e}}_i = \sin\beta$, where β is the angle between $\hat{\mathbf{e}}_i$ and $\mathbf{r}$. For an unpolarized beam, the averaging over the azimuthal angle then gives $[1 - \hat{r}(\hat{r}.)]\hat{\mathbf{e}}_i = (1 + \cos^2\theta)/2$.

Chapter 3

Approximate formulas

3.1 The need for approximate formulas

When none of the rigorous approaches, analytic or numerical, lends itself to an exact solution, one is left with no other choice but to take resort to approximation methods. Sometimes however, the approximation methods are a preferred alternative. This is because (i) they offer simple analytic expressions which are easy to use. Thus, whenever they provide sufficient accuracy, it is desirable to employ them. (ii) The closed form expressions that they provide result in the development of better understanding of the processes involved, and they provide a deeper physical insight into the problem. (iii) Computationally demanding implementations sometimes become constrained by roundoff errors; particularly at large x or where large number of particles of various sizes and shapes are involved. (iv) Modelling a scatterer by a shape itself introduces an approximation. A true description of a shape in terms of a mathematical equation can only be a rare occurrence.

A large number of approximations have been designed over the years for use in specific $m - x$ domains. Related reviews can be found in Mahood (1987), Maslowska (1991), Kim and Lior (1995), Kim et al. (1996), Kokhanovsky and Zege (1997), Wriedt (1998), Wriedt and Comeberg (1998), Sharma and Somerford (1999, 2006), Jones (1999) and Sharma (2013).

3.2 Efficiency factors of small particles

Small particles are identified by the condition $x < 1$. That is, the size of the particle is small in comparison to the wavelength of the radiation. Approximations for this class of spherical scatterers are often based on making a straightforward expansion of Mie scattering coefficients up to certain power of x. Alternately, one may also make approximations by terminating the Mie series, retaining terms up to a certain value of n. A review article, specific to early small particle approximations, is by Kim et al. (1996).

The following expansions for Ricatti–Bessel and Ricatti–Hankel functions are often used in determining small particle approximations.

$$\psi_1(x) = \frac{\sin x}{x} - \cos x \qquad \approx \frac{x^2}{3} - \frac{x^4}{30}, \tag{3.1a}$$

$$\psi_2(x) = \left(\frac{3}{x^2} - 1\right)\sin(x) - \frac{3}{x}\cos(x) \qquad \approx \frac{x^3}{15}, \tag{3.1b}$$

$$\xi_1(x) = e^{ix}\left(-\frac{i}{x} - 1\right) \qquad \approx -\frac{i}{x} - \frac{ix}{2} + \frac{x^2}{3}, \tag{3.1c}$$

$$\xi_2(x) = e^{ix}\left(\frac{-3i}{x^2} - \frac{3}{x} - i\right) \qquad \approx -\frac{i3}{x^2}, \tag{3.1d}$$

$$\psi_1'(x) \approx \frac{2x}{3} - \frac{2x^3}{15}, \qquad \psi_2'(x) \approx \frac{x^2}{5}, \tag{3.1e}$$

$$\xi_1'(x) \approx \frac{i}{x^2} - \frac{i}{2} + \frac{2x}{3}, \qquad \xi_2'(x) \approx \frac{i6}{x^3}. \tag{3.1f}$$

These permit a_1, b_1 and a_2 to be written correctly up to order x^6 as below:

$$a_1 = -\frac{2ix^3}{3}\frac{m^2-1}{m^2+2} - \frac{2ix^5}{5}\frac{(m^2-1)(m^2-2)}{(m^2+2)^2} + \frac{4}{9}\frac{(m^2-1)^2}{(m^2+2)^2}x^6, \tag{3.2a}$$

$$a_2 = -\frac{ix^5}{15(2m^2+3)}(m^2-1), \tag{3.2b}$$

and

$$b_1 = -\frac{ix^5}{45}(m^2-1) \tag{3.2c}$$

It is assumed that the magnetic permeability of the host medium and the scatterer are the same.

Alternatively, explicit expressions for a_1 and b_1 in terms of m and x can be obtained by writing,

$$a_1 = \frac{p_1}{p_1 + iq_1}, \qquad b_1 = \frac{r_1}{r_1 + is_1}, \tag{3.3a}$$

where

$$p_1 = A\ \sin x - B\ \cos x, \tag{3.3b}$$

$$q_1 = A\ \cos x + B\ \sin x, \tag{3.3c}$$

with,

$$A = \left[\left(\frac{m^2 - 1}{m^2 x^3}\right) \sin mx + m\left(1 - \frac{m^2 - 1}{m^2 x^2}\right) \cos mx\right] \tag{3.3d}$$

and

$$B = \left[\left(1 + \frac{m^2 - 1}{m^2 x^2}\right) \sin mx - \left(\frac{m^2 - 1}{mx}\right) \cos mx\right]. \tag{3.3e}$$

and

$$r_1 = C\ \sin x - D\ \cos x, \tag{3.3f}$$

$$s_1 = [C\ \cos x + \cos mx] \cos x + D\ \sin x, \tag{3.3g}$$

where

$$C = \left(\frac{m^2 - 1}{mx}\right) \sin mx + \cos mx, \tag{3.3h}$$

and

$$D = m\ \sin mx. \tag{3.3i}$$

These expressions for a_1 and b_1 are correct to all orders in x.

3.2.1 Rayleigh approximation

Mathematically, the conditions for the validity of this approximation may be expressed as,

$$x \ll 1, \tag{3.4a}$$

and

$$|m|x \ll 1. \tag{3.4b}$$

Equations (3.4a) and (3.4b) express the requirement that the particle experiences a uniform field throughout its volume. Under these conditions $|b_1| \ll |a_1|$. The efficiency factors for scattering, absorption and backscattering (equations (2.55), (2.56), and (2.64)), approximated to the leading power in x in a_1, assume the following forms:

$$Q_{sca}^{RA} = \frac{8x^4}{3}\left|\frac{m^2 - 1}{m^2 + 2}\right|^2, \tag{3.5a}$$

$$Q_{abs}^{RA} = 4x \ \text{Im}\left[\frac{m^2-1}{m^2+2}\right], \tag{3.5b}$$

and

$$Q_{back}^{RA} = x^4 \left|\frac{m^2-1}{m^2+2}\right|^2. \tag{3.5c}$$

The superscript RA refers to the Rayleigh approximation.

A formula for Q_{abs} for scattering of microwaves by a water bubble with a small shell thickness has been given by Dombrovsky and Ballis (2010) in the framework of the Rayleigh theory:

$$Q_{abs} = 4x\text{Im}\left[\frac{(2m^2+1)(1-m^2)}{(3m^2/\bar{\delta}) + 2(m^2-1)^2}\right], \tag{3.5d}$$

where

$$\bar{\delta} = \delta/a, \tag{3.5e}$$

and δ is the shell thickness. Typical values of bubble size are about 1 mm and the shell thickness is about 5 μm.

By inserting $m = n_r + in_i$ in (3.5a) and (3.5b), explicit analytic dependence on real and imaginary parts of the relative refractive index can be exhibited as (see, for example, Selamet and Arpaci 1989),

$$Q_{sca}^{RA} = \frac{8x^4}{3}\left(1 - \frac{3\mathcal{M}_2}{\mathcal{M}_1}\right), \quad \text{and} \quad Q_{abs}^{RA} = 12x\left(\frac{\mathcal{N}_1}{\mathcal{M}_1}\right). \tag{3.6a}$$

where,

$$\mathcal{N}_1 = 2n_r n_i, \quad \mathcal{N}_2 = n_r^2 - n_i^2, \quad \mathcal{M}_1 = \mathcal{N}_1^2 + (2+\mathcal{N}_2)^2, \quad \mathcal{M}_2 = 1 + 2\mathcal{N}_2. \tag{3.6b}$$

For $x \ll 1$, $Q_{sca}^{RA} \ll Q_{abs}^{RA}$. Thus, the absorption efficiency factor is approximately the same as the extinction efficiency factor. On the other hand, for a nonabsorbing particle, $Q_{sca} = Q_{ext}$ trivially.

The term b_1 also needs to be included in the Rayleigh approximation if the condition $b_1 \ll a_1$ is not satisfied:

$$b_1 = \frac{-2ix^3}{3}\frac{\mu_{rel} - 1}{\mu_{rel} + 2}. \tag{3.7}$$

The relative permeability of the scatterer is μ_{rel}. Note that for $\mu_{rel} = 1$, $b_1 = 0$.

For nonspherical particles, the efficiency factors in the RA can be obtained through standard electrostatic approximation by writing,

$$C_{sca} = \frac{k^4}{6\pi}\,|\alpha_p|^2, \tag{3.8a}$$

and

$$C_{abs} = k \text{ Im } \alpha_p, \tag{3.8b}$$

where α_p is the polarizability in the electrostatic approximation:

$$\alpha_p = \mathcal{V}\left(L + \frac{\varepsilon_0}{\varepsilon_1 - \varepsilon_0}\right)^{-1}. \tag{3.9}$$

In (3.9), ε_0 and ε_1 are the permittivities of the medium and the particle, respectively, $\mathcal{V}$ is the volume of the particle and L is the depolarization factor.

For a spherical particle $L = 1/3$. Then, α_p becomes

$$\alpha_p = 3\mathcal{V}\frac{\varepsilon_1 - \varepsilon_0}{\varepsilon_1 + 2\varepsilon_0} \quad = \quad 4\pi a^3 \frac{(m^2 - 1)}{(m^2 + 2)}. \tag{3.10}$$

Equations (3.8a) and (3.8b), with α_p given by (3.10), lead to (3.5a) and (3.5b).

For metallic particles, the dielectric constant can be negative and wavelength dependent. If the refractive index of the particle is such that $m^2 \approx -2$, the polarizability becomes large. The result is large efficiency factors at the corresponding frequency. These are localized surface plasmon resonances. The applicability of the RA to such particles is restricted to $a \approx 5\ nm$ (Ru et al. 2013). The wavelength range in this study was from 360 nm to 440 nm. To extend the validity of the RA to larger spheres, a higher order expansion has been proposed by Meier and Wokaun (1983). The analytic expression for α_p, arrived at by an expansion of a_1, can be expressed as (Kuwata et al. 2003):

$$\alpha_p \approx \left(1 - \varepsilon_0\left(\frac{1+\varepsilon}{10}\right)x^2 + O(x^4)\right)\mathcal{V}\times$$

$$\left[\left(\frac{\varepsilon+2}{\varepsilon-1}\right) - \frac{1}{30}\varepsilon_0\left(\varepsilon+10\right)x^2 - i\frac{4\pi^2\varepsilon_0^{3/2}}{3}\frac{\mathcal{V}}{\lambda_0^3} + O(x^4)\right]^{-1}, \tag{3.11}$$

where $\varepsilon = \varepsilon_1/\varepsilon_0$. The approximation (3.11) is found to be valid up to $a \approx 20 - 30\ nm$ (Ru et al. 2013, Schebarchov et al. 2013).The numerator as well as the denominator (the inverse term) are exact to the order of x^3. For $\varepsilon_r \gg \varepsilon_i$, the second term in the denominator is real (the surrounding mediums dielectric constant ϵ_0 is real) and is responsible for the shift of the resonance. The third term is the radiation damping correction originating because of the size dependence of the electrostatic approximation.

The range of validity of (3.11) can be extended further by including terms to the order x^4 (Schebarchov et al. 2013). The modified formula gives credible results up to a radius of about 50 nm and in reasonable agreement up to $a \approx 70\ nm$. The contributions of the lowest order terms of the magnetic dipole and electrical quadrupole have also been examined for silver nanospheres. The resulting approximation for far-field properties is then extremely good up to $a \sim 70\ nm$. The contribution of the magnetic dipole term is negligible.

A generalization of (3.11) to nonspherical objects can be achieved by writing (Kuwata et al. 2003):

$$\frac{\alpha_p}{\mathcal{V}} = \left[\left(L + \frac{1}{\varepsilon - 1}\right) + A\varepsilon_0 x^2 + B\varepsilon_0^2 x^4 - i\frac{4\pi^2\varepsilon_0^{3/2}}{3}\frac{\mathcal{V}}{\lambda_0^3}\right]^{-1}. \tag{3.12}$$

For $\varepsilon_r \gg \varepsilon_i$ and low radiation loss, the resonant condition,

$$\left(L + \frac{1}{\varepsilon - 1}\right) + A\varepsilon_0 x^2 + B\varepsilon_0^2 x^4 = 0, \tag{3.13}$$

allows estimation of A and B using light scattering spectra. For a prolate spheroid this gives,

$$A(L) = -0.4865L - 1.046L^2 + 0.8481L^3, \tag{3.14}$$

and

$$B(L) = 0.01909L + 0.1999L^2 + 0.6077L^3, \tag{3.15}$$

which are materially independent.

Closed form expressions have been obtained for nanoshells (Alam and Massoud 2006, Li et al. 2009). In the later reference, expressions for a_1, a_2 and b_1 have been obtained up to the power $\bar{y}^6$, where $\bar{y} = x + t$ with $0.01 \le t \le 0.4\ nm$ as the shell thickness. A numerical comparison of exact light scattering characteristics of particles in the size range $0.02 \le \bar{y} \le 1.4$ has been made by the predictions of Alam and Massoud (2006). The approximation is found to be very accurate over a large range of parameters of interest including those in optical nanotechnology. The coefficient a_1 is the main contribution to the cross-sections. The contribution of coefficients b_1 and a_2, although small, improve the accuracy of extinction and scattering cross sections.

Employing (3.5a) and (3.5b) in (1.68), one can write the single scattering albedo in the RA as (Brown 2013),

$$\omega_0 = 1 - \frac{9n_r n_i}{x^3[(n_r^2 + n_i^2)^2 - 2(n^2 - n_i^2) + 1] + 9n_r n_i}, \tag{3.16}$$

and the variation of the single scattering albedo with x can be expressed in a simple form:

$$\frac{d\omega_0}{dx} = \frac{x^2[27n_r n_i((n_r - 1)^2 + n_i^2)((n_r + 1)^2 + n_i^2)]}{[x^3((n_r^2 + n_i^2)^2 - 2(n_r^2 - n_i^2) + 1) + 9n_r n_i]^2}. \tag{3.17}$$

In the limiting case $x \to 0$, the above expression can be shown to reduce to,

$$\frac{d\omega_0}{dx} = \frac{x^2[(n_r-1)^2+n_i^2][(n_r+1)^2+n_i^2]}{3n_r n_i}. \tag{3.18}$$

These equations have been used in connection with studies on regions of spectral blueing on planetary surfaces.

3.2.2 The Tien–Doornink–Rafferty approximation

This approximation gives absorption efficiency as (Tien et al. 1972):

$$Q_{abs}^{TDRA} = 2\left[1-e^{-G\rho}\right], \tag{3.19}$$

where

$$G = \frac{6n_r}{4n_r^4-8n_r^3+8n_r^2+4n_r+1}, \tag{3.19a}$$

and $\rho = 2x(n_r-1)$.

3.2.3 The first-term approximation

The first-term approximation (FTA) retains coefficients a_1 and b_1 in the Mie series in their exact form. Thus,

$$Q_{ext}^{FTA} = \frac{6}{x^2}\,\mathrm{Re}\Big(a_1+b_1\Big), \tag{3.20a}$$

and

$$Q_{sca}^{FTA} = \frac{6}{x^2}\,\Big(|a_1|^2+|b_1|^2\Big), \tag{3.20b}$$

constitute the first-term approximation.

3.2.4 Wiscombe approximation

Wiscombe approximation (WA) (Wiscombe 1980) retains three Mie coefficients, namely, a_1, b_1 and a_2. These are scaled by a factor $1/x^3$ for computational convenience. The scaled coefficients are denoted as $\hat{a}_1 = a_1/x^3$, $\hat{a}_2 = a_2/x^3,\ and\hat{b}_1 = b_1/x^3$. The extinction and scattering efficiency factors (2.55) and (2.56) can then be cast as,

$$Q_{ext}^{WA} = 6x\ \mathrm{Re}\Big(\hat{a}_1+\hat{b}_1+\frac{5}{3}\hat{a}_2\Big), \tag{3.21}$$

and

$$Q_{sca}^{WA} = 6x^4\Big(|\hat{a}_1|^2+|\hat{b}_1|^2+\frac{5}{3}|\hat{a}_2|^2\Big), \tag{3.22}$$

where the approximations for the scaled Mie coefficients are,

$$\hat{a}_1 = \frac{2i(m^2-1)}{3D}\left(1-\frac{x^2}{10}+\frac{(4m^2+5)}{1400}x^4\right), \tag{3.23a}$$

$$\hat{b}_1 = \frac{ix^2(m^2-1)}{45}\left(1+\frac{(2m^2-5)x^2}{70}\right)\left(1-\frac{(2m^2-5)x^2}{30}\right)^{-1}, \qquad (3.23b)$$

$$\hat{a}_2 = ix^2(m^2-1)\left(1-\frac{x^2}{14}\right)\left[15\left(2m^2+3-\frac{(2m^2-7)x^2}{14}\right)\right]^{-1}, \qquad (3.23c)$$

with

$$D = (m^2+2)+\left(1-\frac{7m^2}{10}\right)x^2-\frac{(8m^4-385m^2+350)x^4}{1400}$$
$$+\frac{2i(m^2-1)}{3}\left(1-\frac{x^2}{10}\right)x^3. \qquad (3.23d)$$

While arriving at these expressions the van de Hulst (1957) convention has been followed. These equations can be converted to the present notation by taking their complex conjugate.

3.2.5 Penndorf approximation

Penndorf (1962) derived an approximate expansion by truncating terms of order x^5 and higher for Q_{ext} and x^8 and higher for Q_{sca}. The approximate extinction and scattering efficiency factors work out to be:

$$Q_{ext}^{PAPP} = \frac{24n_r n_i x}{z_1}+\left[\frac{4}{15}+\frac{20}{3z_2}+\frac{48}{z_1^2}\left(7(n_r^2+n_i^2)+4(n_r^2-n_i^2-5)\right)\right]n_r n_i x^3$$
$$+\frac{8}{3z_1^2}\left[\left((n_r^2+n_i^2)^2+n_r^2-n_i^2-2\right)^2-36n_r^2n_i^2\right]x^4, \qquad (3.24a)$$

and

$$Q_{sca}^{PAPP} = \frac{8x^4}{3z_1^2}\left[\left((n_r^2+n_i^2)^2+n_r^2-n_i^2-2\right)^2-36n_r^2n_i^2\right]x^4$$
$$\times\left[1+\frac{6}{5z_1}\left((n_r^2-n_i^2)^2-4\right)x^2-\frac{8n_r n_i x^3}{z_1}\right], \qquad (3.24b)$$

respectively. The superscript $PAPP$ stands for Penndorf approximation and

$$z_1 = (n_r^2+n_i^2)^2+(n_r^2-n_i^2)+4, \qquad (3.25)$$

$$z_2 = 4(n_r^2+n_i^2)^2+12(n_r^2-n_i^2)+9. \qquad (3.26)$$

A more compact form for the efficiency factors is (Selamet and Arpaci 1989):

$$Q_{ext}^{PAPP} = 2x^3\left[\mathcal{N}_1\left(\frac{1}{15}+\frac{5}{3\mathcal{M}_4}+\frac{6\mathcal{M}_5}{5\mathcal{M}_1^2}\right)+\frac{4\mathcal{M}_6}{3\mathcal{M}_1^3}x\right], \qquad (3.27a)$$

and

$$Q_{sca}^{PAPP} = Q_{sca}^{RA}\left[1+2\frac{x^2}{\mathcal{M}_1}(0.6\mathcal{M}_3-2\mathcal{N}_1 x)\right], \qquad (3.27b)$$

where

$$\mathcal{M}_3 = \mathcal{N}_3 - 4, \quad \mathcal{M}_4 = 4\mathcal{N}_1^2 + (3 + 2\mathcal{N}_2)^2, \quad \mathcal{M}_5 = 4(_2 - 5) + 7\mathcal{N}_3,$$

$$\mathcal{M}_6 = (\mathcal{N}_2 + \mathcal{N}_3 - 2)^2 - 9\mathcal{N}_1^2, \quad \text{and} \quad \mathcal{N}_3 = \mathcal{N}_1^2 + \mathcal{N}_2^2. \tag{3.27c}$$

3.2.6 Caldas–Semião approximation

This approximation retains terms up to the order of x^6 in the expansion for Q_{ext} and up to the order of x^9 in the expansion for Q_{sca}. The efficiency factors and the asymmetry parameter so obtained are (Caldas and Semião 2001a):

$$Q_{ext}^{CSA} = x\left[-4\ \mathrm{Im}(P) - \frac{2}{15}\ \mathrm{Im}\left[P\left(Q' + V' + S'\right)\right]\ x^2 + \frac{8}{3}\ \mathrm{Re}(P^2)\ x^3\right.$$

$$\left. -\frac{2}{1575}\ \mathrm{Im}\left[P\left(R' + W' + T' + 5V' - U'\right)\right]\ x^4 + \frac{8}{45}\ \mathrm{Re}(P^2Q')\ x^5\right], \tag{3.28a}$$

$$Q_{sca}^{CSA} = \frac{8}{3}|P|^2\ x^4\left[1 + \frac{1}{15}\ \mathrm{Re}(Q')\ x^2 + \frac{4}{3}\ \mathrm{Im}(P)\ x^3 + \right.$$

$$\left.\frac{1}{31500}\left(35|Q'|^2 + 20\ \mathrm{Re}(R') + 35|V'|^2 + 21|S'|^2\right)\ x^4 + \frac{2}{45}\ \mathrm{Im}\left(Q'(P - \bar{P})\right)\ x^5\right], \tag{3.28b}$$

$$Q_{abs}^{CSA} = x\left[-4\ \mathrm{Im}(P) - \frac{2}{15}\ \mathrm{Im}\left[P\left(Q' + V' + S'\right)\right]\ x^2 + \frac{16}{3}\ \mathrm{Im}(P^2)\ x^3\right.$$

$$\left. -\frac{2}{1575}\ \mathrm{Im}\left[P\left(R' + W' + T' + 5V' - U'\right)\right]\ x^4 + \frac{8}{45}\ \mathrm{Im}(P)\mathrm{Im}(PQ')\ x^5\right], \tag{3.28c}$$

and

$$g^{CSA} = \frac{4}{225\ Q_{sca}}|P|^2x^6\left[(5\ \mathrm{Re}(V') + 3\ \mathrm{Re}(S')) + \right.$$

$$\frac{1}{210}\left(35\ \mathrm{Re}(V'Q'^*) + 21\ \mathrm{Re}(S'Q'^*) + 10\ \mathrm{Re}(W') - 6\ \mathrm{Re}(T')\right)\ x^2$$

$$\left. -\frac{2}{3}\left(5\ \mathrm{Im}(V'P^*) - 3\ \mathrm{Im}(S'P^*)\right)\ x^3\right], \tag{3.28d}$$

where

$$P = \frac{m^2 - 1}{m^2 + 2}, \qquad Q = \frac{m^2 - 2}{m^2 + 2},$$

$$R = \frac{m^6 + 20m^4 - 200m^2 + 200}{(m^2+2)^2}, \qquad S = \frac{m^2-1}{2m^2+3},$$

$$T = \frac{m^2-1}{(2m^2+3)^2}, \qquad U = \frac{m^2-1}{3m^2+4}, \qquad V = m^2-1, \qquad W = (m^2-1)(2m^2-5),$$

$$Q' = Q, \qquad R' = 18R, \qquad S' = \frac{5S}{P}, \qquad T' = \frac{375T}{P},$$

$$U' = \frac{28U}{P}, \qquad V' = \frac{V}{P}, \qquad W' = \frac{5W}{P}. \tag{3.28e}$$

As usual, the asterisk over a symbol represents the complex conjugate of the quantity.

3.2.7 Numerical comparisons

Error in the RA has been estimated by many researchers. Walstra (1964) and Heller (1963) presented error contour charts in the $m - x$ domain for nonabsorbing spheres. Penndorf (1962) examined error contour charts for absorbing ($n_r = 1.29$, $0.0645 \leq n_i \leq 5.16$ and $n_r = 1.25$, 1.50, 1.75; $1.05 \leq n_i \leq 1.25$) as well as nonabsorbing ($1.05 \leq n_r \leq 2.0$) spheres. Ku and Felske (1984) give a number of simple analytic criteria for ensuring an error of $\leq 1\%$ in scattering and extinction efficiencies. The range of refractive index values considered is $1.01 \leq n_r \leq 50$; $0 \leq n_i \leq 50$. Selamet and Arpaci (1989) examined the validity of the RA for $m = 1.5 + i0.5$, $2.0 + i1.0$, $1.75 + i0.75$. The error was less than 10% provided $x \leq 0.3$. The x limit increases to $x \leq 0.8$ for the Penndorf approximation.

Li et al. (2015) examined the validity of the RA for precipitation particles of fixed $x = 0.13$ for $\lambda(cm) = 10, 5, 3, 2.2$. The errors in the RA ranged from 9% for $\lambda = 10\ cm$ to about 5% for $\lambda = 2.2\ cm$. It was concluded that conditions (3.4a) and (3.4b) constitute sufficient but not necessary conditions for the validity of the RA.

The Wiscombe approximation is accurate up to six significant digits if $x \leq 0.1$; 4-5 digits if $x \leq 0.2$. Both the accuracy limits further require $|m| \leq 2$. Its accuracy decreases as $|m|$ increases and is not recommended for a more than six-digit accuracy if $|m|x \leq 0.1$ (Wiscombe 1980).

Kim et al. (1996) have compared the radiative properties predicted by the RA, the PA, the WA, and the FTA against Mie theory computations in the relative refractive index range $1.0 \leq n_r \leq 5.0$ and $0.001 \leq n_i \leq 50$ for $0.0 \leq x \leq 1.0$. It was concluded that while the choice of approximation depends on the refractive index and the size parameter, the first-term approximation is best in most cases.

The Caldas–Semião approximation (CSA) has been tested against the Mie theory and the PAPP for $0 \leq x \leq 1$, $1 \leq n_r \leq 5$ and $10^{-6} \leq n_i \leq 4$ (Caldas-Semião 2001a). Comparisons show that the CSA constitutes significant improvement over the PAPP for efficiency factors as well as for asymmetry parameters.

3.2.8 Videen and Bickel approximation

When the condition $|m|x \ll 1$ is not satisfied, the expansion in powers of $|m|x$ is not valid. Thus, Videen and Bickel (1992) proposed an approximation for real values of m, where the first two Mie coefficients are retained and expanded in powers of x up to $O(x^3)$. The expansion differs from the expansion in (3.2) in that the sine and cosine functions are left untouched. The expansion coefficients a_1 and b_1 are thus written as:

$$a_1 \approx \frac{A_{11}\cos mx + A_{12}\sin mx}{A_{13}\cos mx + A_{14}\sin mx}, \tag{3.29}$$

and

$$b_1 \approx \frac{B_{11}\cos mx + B_{12}\sin mx}{B_{21}\cos mx + B_{22}\sin mx}, \tag{3.30}$$

where

$$A_{11} = \frac{(1+2m^2)x}{3m} - \frac{(1+4m^2)x^3}{30m}, \quad A_{12} = -\frac{1+2m^2}{3m^2} + \frac{(1+14m^2)x^2}{30m^2},$$

$$A_{13} = \frac{-i+im^2}{mx^2} - \frac{i+im^2}{2m}, \quad A_{14} = \frac{i-im^2}{m^2x^3} + \frac{i-im^2}{2m^2x}, \tag{3.31}$$

and

$$B_{11} = \left(x - \frac{x^3}{6}\right), \qquad B_{12} = \left(\frac{-1}{m} + \frac{(1+2m^2)x^2}{6m}\right),$$

$$B_{21} = (-i+x), \qquad B_{22} = \left(\frac{i-im^2}{mx} - \frac{1}{m} - \frac{i(1+m^2)x}{2m}\right). \tag{3.32}$$

The leading term in the limit $x \to 0$ arises from a_1 and yields the RA. On the other hand, as $m \to i\infty$, the scattering coefficients reduce to $a_1 \sim -(2ix^3/3)$ and $b_1 \sim (ix^3/3)$. In both limiting cases the scattering coefficients are proportional to x^3.

The expressions for a_2 and b_2 obtained are,

$$a_2 \approx \frac{A_{21}\cos mx + A_{22}\sin mx}{A_{23}\cos mx + A_{24}\sin mx}, \tag{3.33}$$

and

$$b_2 \approx \frac{B_{21}\cos mx + B_{22}\sin mx}{B_{23}\cos mx + B_{24}\sin mx}, \tag{3.34}$$

where

$$A_{21} = -\left(\frac{2+3m^2}{5m^2}\right)x + \left(\frac{6+29m^2}{210m^2}\right)x^3,$$

$$A_{22} = \left(\frac{2+3m^2}{5m^2}\right) - \left(\frac{2+19m^2+14m^4}{70m^3}\right)x^2,$$

$$A_{23} = \left(\frac{18i-18im^2}{m^2x^4}\right) + \left(\frac{3i-im^2}{m^2x^2}\right),$$

$$A_{24} = \left(\frac{18i-18im^2}{-m^3x^5}\right) + \left(\frac{-3i+9im^2-6im^4}{m^3x^3}\right), \tag{3.35}$$

and

$$B_{21} = \frac{-x}{m} + \frac{(3+2m^2)x^3}{30m}, \qquad B_{22} = \frac{1}{m^2} - \frac{(1+4m^2)x^2}{10m^2},$$

$$B_{23} = \frac{3i(1-m^2)}{mx^2} + \frac{i(3-m^2)}{2m}, \qquad B_{24} = \frac{3i(m^2-1)}{m^2x^3} + \frac{3i(m^2-1)}{2m^2x}. \tag{3.35a}$$

The extension of the Videen and Bickel formula to the case of relative permeability, $\mu \neq 1$, is straightforward. The Mie coefficients a_1 and a_2 are then formally identical to (3.29) and (3.33), but with the coefficients of sine and cosine functions altered (Garcia-Camara et al. 2008). The modified coefficients are:

$$A_{11} = m^2\tilde{m}x^4, \qquad A_{12} = -m\tilde{m}x^3,$$

$$A_{13} = mx(-m\tilde{m}x^3 - im\tilde{m} + ix^2 + i),$$

and

$$A_{14} = (m\tilde{m}x^3 + im\tilde{m} - ix^2 + im^2x^4 - i + im^2x^2). \tag{3.36}$$

where $\tilde{m} = m/\mu$. The relevant coefficients for a_2 are:

$$A_{21} = mx(6m\tilde{m}x - 6x + m^2x^3),$$

$$A_{22} = x(6 - 3m^2x^2 - 6m\tilde{m} + 2m^3\tilde{m}x^2),$$

$$A_{23} = 3im^3\tilde{m}x^2 + 2m^3\tilde{m}x^3 - im^2\tilde{m}x^4 + 6im^3\tilde{m} - 9im\tilde{m} - 6m\tilde{m}x$$
$$+3im\tilde{m}x^2 + 18\frac{im\tilde{m}}{x^2} - \left(-\frac{3i}{x^2} - 2i - x\right)(6 - 3m^2x^2),$$

$$A_{24} = 9im^2\tilde{m}x + 6m^2\tilde{m}x^2 - 3im^2\tilde{m}x^3$$
$$+18\frac{im^2\tilde{m}}{x} - \left(-\frac{3i}{x^2} - 2i - x\right)(-6mx + 3m^3x^3), \tag{3.36a}$$

and

$$b_n\left(\frac{1}{\tilde{m}}, m, x\right) = a_n(\tilde{m}, m, x), \tag{3.37}$$

because of the symmetry of the scattering coefficients.

If the sine and cosine functions are also expanded in powers of x, and the terms are up to order x^3 in the numerator and denominator of a_1 and the terms are up to x^5 in the numerator and denominator of a_2, one obtains (Ambrosio 2015):

$$a_1 = -\frac{2}{3}\frac{(1-\varepsilon)x^3}{i(\varepsilon+2) - i\frac{\left(-10+4m^2+\varepsilon(m^2+5)\right)x^2}{10} - 2(1-\varepsilon)x^3/3}, \tag{3.38}$$

and

$$a_2 = -\frac{1}{15}\frac{(1-\varepsilon)x^5}{i(2\varepsilon+3) - ifx^2 + igx^4 - \frac{(1-\varepsilon)x^5}{15}}, \tag{3.39}$$

where,

$$f = \frac{-7 + m^2(2\varepsilon+5)}{14}, \tag{3.40}$$

and

$$g = \frac{63 - 30m^2 + 7m^2 + 2\varepsilon(m^4 - 21)}{504}. \tag{3.41}$$

If $\varepsilon = -2$, the first term in the denominator of a_1 is zero. But the second and third terms ensure that the denominator is not zero. Further, if $\varepsilon = -2$ and $\mu = -5$ ($m^2 = \varepsilon\mu$), then the third term ensures that $a_1 = 1$ and $Q_{ext} \sim 1/x^2$. Notice, that it is different from the usual $1/x^4$ behaviour of the extinction in the RA (Miroshnicchenko 2009, Ambrosio 2015).

It is clear from the above that the first resonance does not occur exactly at $\varepsilon = -2$. In fact, it can be shown from (3.29) that the ε for which a_1 becomes the maximum is given by the expression:

$$\varepsilon = \left(\frac{1}{6\mu x^2}\right)\Bigg[30 - 15x^2 - 12\mu x^2 + 20ix^3 - i\Big(-900(1-x^2) - 1200ix^3 - 225x^4$$

$$-720\mu x^4 - 144\mu^2x^4 + 600ix^5 + 720i\mu x^5 + 400x^6\Big]^{1/2}\Bigg]. \tag{3.42}$$

Clearly, a_1 is maximum at $\varepsilon = -2$ only at $x = 0$.

One may also write,

$$g = \frac{4}{Q_{sca}x^2}\left[\frac{3}{2}\mathrm{Re}(a_1a_2^* + b_1b_2^* + a_1b_1^*) + \frac{8}{3}\mathrm{Re}(a_2a_3^*) + \frac{5}{6}\mathrm{Re}(a_2b_2^*)\right], \tag{3.43}$$

by retaining a_1, a_2, a_3, b_1, b_2 in the Mie series expansion (Dombrovsky and Baillis 2010).

3.3 Angular scattering by small particles: Parameterization

3.3.1 Five-parameter phase function

A possible way of parameterizing a given phase function, displaying only one minimum, is to express it in terms of an expansion in powers of $\cos\theta$ (Sharma et al. 1998):

$$\phi_{FPPF} = b_0 + b_1\cos\theta + b_2\cos^2\theta + b_3\cos^3\theta + b_4\cos^4\theta, \tag{3.44}$$

where, the letters $FPPF$ in the subscript refers to a five-parameter phase function. The coefficients $b_0, ...b_4$ can be determined by imposing the following constraints:

1. The ϕ_{FPPF} is normalized according to the relation

$$2\pi\int_{-1}^{1}\phi_{FPPF}(\cos\theta)d(\cos\theta) = 1. \tag{3.45}$$

2. The phase functions ϕ_{FPPF} matches ϕ_{EX} at $\theta = 0, \pi$ and $\pi/2$.

3. The slope of ϕ_{FPPF} is identical with that of ϕ_{EX} at $\theta = \pi/2$.

The letters in the subscript EX denote the exact (to be parameterized) phase function. The implementation of the constraints (1–3) leads to the following simple expressions for the coefficients:

$$b_0 = \phi_{EX}(\pi/2), \tag{3.46a}$$

$$b_1 = \left.\frac{d\phi_{EX}}{d(\cos\theta)}\right|_{\theta=\pi/2}, \tag{3.46b}$$

$$b_2 = \frac{3}{8\pi}\left[5 - 2\pi\Big(\phi_{EX}(0) + \phi_{EX}(\pi) + 8\phi_{EX}(\pi/2)\Big)\right], \tag{3.46c}$$

$$b_3 = \frac{\phi_{EX}(0) - \phi_{EX}(\pi)}{2} - a_1, \tag{3.46d}$$

$$b_4 = \frac{5}{8\pi}\left[2\pi\Big(\phi_{EX}(0) + \phi_{EX}(\pi) + 4\phi_{EX}(\pi/2)\Big) - 3\right]. \tag{3.46e}$$

A numerical evaluation of accuracy of (3.44) has been done by generating exact phase functions for several values of size parameter and refractive indices for spheres (Sharma et al. 1998) and for infinitely long cylinders and spheroids (Sharma and Roy 2000). The formula gives accurate results if $x \leq 2$ and if $m \leq 1.5$. Figure 3.5 shows a comparison of the exact phase function and its parametrized form (3.44) for $m = 1.15$ and $x = 0.1$, 0.5 and 1.0. The agreement is good even for a refractive index of $m = 3.0$ if $x \leq 1.0$. The result is independent of the shape of the particle as long as $x \leq 1$, $g \leq 0.6$ and the phase function does not have more than one minimum.

A simple analytic expression can also be obtained for the asymmetry parameter from (1.71):

$$g_{FPPF} = \frac{2\pi}{15}\left[4a_1 + 3[\phi_{mie}(0) - \phi_{mie}(\pi)]\right]. \tag{3.47}$$

Equation (3.47) allows one to obtain the value of the asymmetry parameter in a straightforward way. Numerical tests show it to be in excellent agreement with the exact results.

Alternatively, (3.44) may be expressed as

$$\phi(\cos\theta)_{FPPF} = a_0 P_0(\cos\theta) + a_1 P_1(\cos\theta) + a_2 P_2(\cos\theta) + a_3 P_3(\cos\theta) + a_4 P_4(\cos\theta). \tag{3.48}$$

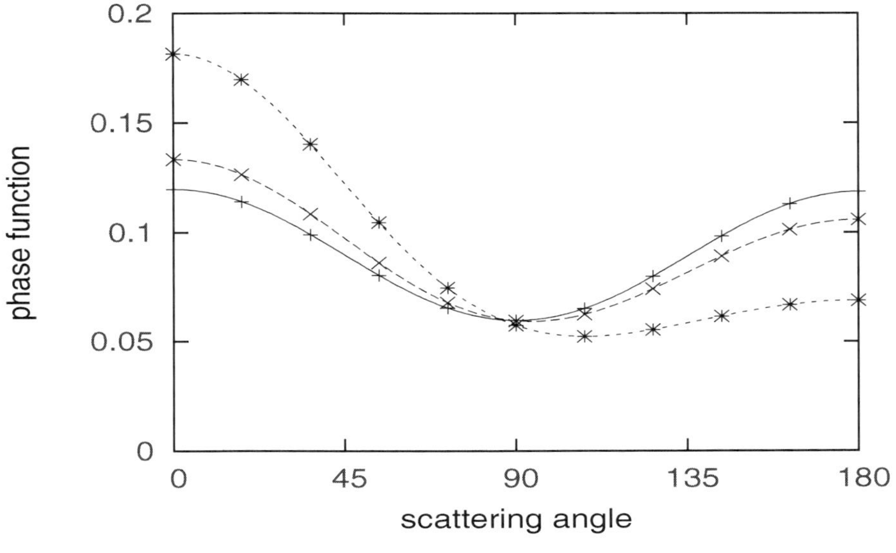

FIGURE 3.1: Comparison of parameterizations obtained using (3.44) with exact sphere phase functions of $x = 0.1$, 0.5 and 1.0. The relative refractive index in all three cases is 1.15. Solid line: $x = 0.1$, large dashed line: $x = 0.5$ and small dashed line: $x = 1.0$. The continuous lines are exact results and the points are result of parameterization. This figure was first published in Sharma et al. (1998)

This is equivalent to (3.44) with the following relations between coefficients:

$$a_0 = b_0 + \frac{1}{3}b_3 + \frac{1}{5}b_5, \tag{3.48a}$$

$$a_1 = b_1 + \frac{3}{5}b_3, \tag{3.48b}$$

$$a_2 = \frac{2}{3}b_2 + \frac{4}{7}b_4, \tag{3.48c}$$

$$a_3 = \frac{2}{5}b_3, \tag{3.48d}$$

$$a_4 = \frac{8}{35}b_4. \tag{3.48e}$$

The values of a_l coefficients so derived agree well with those obtained from,

$$a_l = \frac{2l+1}{2}\int_{-1}^{1} \phi_{EX}(\mu)P_l(\mu)\, d\mu. \tag{3.49}$$

This was verified for two types of particles (Sharma et al. 1999): (i) absorbing spheres of $x = 0.7854$ and $m = 1.42 - i0.05$ and (ii) nonabsorbing spheres with $x = 1.48$ and $m = 1.818$. These are titanium dioxide pigment particles widely used in the paint industry to mix with paint for dispersion of dark pigments.

3.3.2 Six-parameter phase function

Improvement in the five–parameter phase function can be obtained by adding one more term in (3.44). The additional parameter may be determined by an additional condition requiring that the phase function reproduces the correct value of g as well. This additional constraint leads to the following phase function:

$$\phi(\cos\theta)_{SIXPPF} = b'_0 + b'_1\cos\theta + b'_2\cos^2\theta + b'_3\cos^3\theta$$
$$+ b'_4\cos^4\theta + b'_5\cos^5\theta. \tag{3.50}$$

where $b'_0 = b_0,\ b'_1 = b_1,\ b'_2 = b_2,\ b'_4 = b_4$ with

$$b'_3 = \frac{35 g_{EX}}{8\pi} - \frac{10b_1}{3} - \frac{5}{4}\left[\phi_{EX}(0) - \phi_{EX}(\pi)\right], \tag{3.50a}$$

and

$$b'_5 = -\frac{35 g_{EX}}{8\pi} + \frac{7b_1}{3} + \frac{7}{4}\left[\phi_{EX}(0) - \phi_{EX}(\pi)\right], \tag{3.50b}$$

The above phase function agrees well with the exact phase function even for $x = 2.0$, $m = 1.5$. The corresponding asymmetry factor is $g = 0.62$. The subscript $SIXPPF$ indicates a six-parameter phase function.

3.3.3 Series expansion

It was suggested long ago by Hartel (1940) that the Mie phase function can be expressed in the form,

$$\phi_{MIE}(\mu) = \frac{1}{4\pi}\sum_{l=0}^{n-1} a_l P_l(\mu), \tag{3.51}$$

where $P_l(\mu)$ is the l-th order Legendre polynomial and a_l are the expansion coefficients. The expansion coefficients a_l can be expressed as,

$$a_l = \frac{2l+1}{2}\int_{-1}^{1}(4\pi)\phi(\mu)P_l(\mu)d\mu. \tag{3.52}$$

For unpolarized wave scattering, phase function expansion coefficients have been given by Chu and Churchill (1955), Clark et al. (1957) and Allen (1974). Equation (3.51) can also be expanded as a series in powers of μ (Pegoraro et al. 2010),

$$\phi_{MIE}(\mu) = \sum_{n=0}^{n-1} c(n)\mu^n. \tag{3.53}$$

The coefficients $c(n)$ are related to the coefficients a_n via the relation:

$$c(n) = \frac{1}{4\pi}\sum_{k=n}^{\leq N-1} a_k \frac{(-1)^{\frac{k-n}{2}}}{2^k}\frac{(k+n)!}{(\frac{k+n}{2})!(\frac{k-n}{2})!n!}. \tag{3.54}$$

The expansion coefficients computed in this manner are for unpolarized Mie scattering. A methodology for calculating expansion coefficients of polarized Mie phase functions has been developed by Fowler (1983) from where the first few analytic expansion coefficients for small particles may be computed. A computer code to calculate the phase function for a spherical particle to the desired order of moments has been developed by Edwards and Slingo (1996).

For small particles ($x \ll 1$), the phase function may be approximated by the first few terms of the expansion (3.51). It is easy to see that the zero-th moment of the phase function, a_0, is always 1 because of the normalization condition (3.45). The first term of the expansion (3.51) then gives an isotropic phase function,

$$\phi(\mu) = \frac{1}{4\pi}. \tag{3.55}$$

Next, the first moment of the phase function is,

$$a_1 = \frac{3}{2}\int_{-1}^{1}\cos\theta\phi(\cos\theta)d\cos\theta \quad = 3\langle\cos\theta\rangle, \tag{3.56}$$

where, by definition, $\langle\cos\theta\rangle$ is the asymmetry parameter. Retaining the first two terms in the expansion (3.51) gives:

$$\phi(\cos\theta) = \frac{1}{4\pi}\left[1 + 3\langle\cos\theta\rangle\cos\theta\right]. \tag{3.57}$$

This phase function is referred to as the Eddington phase function.

A modified Eddington phase function,

$$\phi(\mu) = \frac{1}{4\pi}\left[2f\delta(1-\mu) + (1-f)(1+3\langle\mu\rangle\mu)\right], \tag{3.58}$$

is frequently used to represent atmospheric aerosol and tissue phase phase functions. In (3.58), $0 \le f \le 1$ and δ is the delta function. As $f \to 1$ the phase function becomes a delta function and as $f \to 0$ the phase function reduces to the Eddington phase function.

For sea water, McCormick (1987) expresses the phase function as,

$$\phi(\mu) = \frac{1}{4\pi}\left[2f\delta(1-\mu) + (1-f)[1 + 3\langle\mu\rangle\mu + 5kP_2(\mu)]\right], \tag{3.59}$$

where k is the higher order function in comparison to g. Another modified form is by McCormick and Rinaldi (1989):

$$\phi(\mu) = \frac{1}{4\pi}\left[2f\delta(1-\mu) + (1-f)\sum_{n=0}^{\infty} a_n P_n(\mu)\right]. \tag{3.60}$$

Equations for estimating the unknown coefficients have been given by the authors.

A power series for the Mie scattering, with $\sin^2(\theta/2)$ as the expansion parameter, has been developed by Box (1983). This series expansion should prove to be useful whenever the diffraction peak is of primary interest.

3.4 Angular scattering by small particles: Dependence on particle characteristics

The phase functions discussed so far are such that they do not reflect the physical characteristics of the scatterer. These phase functions only give a mathematical form to a given curve. In practical applications, however, it is desirable to have a parametrized phase function showing explicit dependence on m and x. We now consider such phase functions.

3.4.1 Rayleigh phase function

The Rayleigh phase function (RPF) (see, for example, Chandrasekhar 1960) is,

$$\phi_{RPF}(\mu) = \frac{3}{16\pi}\left(1+\mu^2\right). \tag{3.61}$$

The approximation has been modified to,

$$\phi_{RPF}(\mu) = \frac{3}{16\pi(1+2\gamma)}\left[(1+3\gamma)+(1-\gamma)\mu^2\right]. \tag{3.62}$$

where γ accounts for the effect of molecular anisotropy on the Rayleigh scattering. For isotropic scattering, $\gamma = 0$. Equation (3.62) then reduces to (3.61).

The first three coefficients and Legendre polynomials in (3.51) are,

$$a_0 = 1,\ a_1 = 0,\ a_2 = 1/2 : \ \ (P_0(\mu) = 1,\ P_1(\mu) = \mu,\ P_2(\mu) = (3\mu^2 - 1)/2). \tag{3.62a}$$

It may be easily verified that the Rayleigh phase function (3.61) is equivalent to retaining the first three terms in (3.51). This phase function has been found to match reasonably well with the Mie results for small particles ($x \ll 1$).

3.4.2 Phase function for small spherical particles

Roy and Sharma (2008) successfully expressed the Mie phase function at $\theta = 0$, $\theta = \pi/2$, $\theta = \pi$ and the parameter g as equations involving m and x. The following empirical power series expansions were obtained:

$$\phi_{MIE}(0) \approx 0.1187 + 0.0045mx + \left(0.0356 - 0.0005p + 0.005p^2\right)x^2 + \left(0.0032 + 0.008p\right)x^3 + \frac{(3m-3.02)(m+1)}{200m}x^4, \tag{3.63a}$$

$$\phi_{MIE}(\pi/2) \approx 0.0597 - 0.000qx^2 + 0.00125qx^3 - 0.0005\left(4m^2+1\right)x^4, \tag{3.63b}$$

$$\phi_{MIE}(\pi) \approx 0.1209 - 0.007mx - (0.038 + 0.0028p^2)x^2 - \frac{0.0112p^3}{m}x^3 - 0.00369mp^2x^4, \tag{3.63c}$$

$$g_{MIE} \approx -0.01235px + \left(0.1621m - 0.0282p + 0.0561p^2\right)x^2 + \left(\frac{0.0033}{m} - (0.126+0.06m)p\right)x^3 + \left(\frac{0.0033}{m^2} + 0.088mp + 0.004p^4\right)x^4, \tag{3.63d}$$

where $p = (m-1)$ and $q = (2m^2-1)$. Equations (3.63) are valid in the domain $x \leq 1.25$ and $m \leq 2.0$. The typical accuracy of the empirical formulas can be seen in Figure 3.2, which compares prediction of (3.63c) against the Mie computations.

Substituting the expressions (3.63a) to (3.63c) in (3.44), the ϕ_{FPPF} for a sphere can be expressed as a power series in x with θ and m dependent coefficients:

$$\bar{\phi}_{FPPF}(\theta) = A_0 + A_1 x + A_2 x^2 + A_3 x^3 + A_4 x^4, \tag{3.64}$$

where $A_0,\ A_1, ...A_4$ are:

$$A_0 = b_0 + b_1\alpha_1 + b_2\alpha_2 + b_3\alpha_3 + b_4\alpha_4, \tag{3.64a}$$

$$A_1 = b_1\beta_1 + b_2\beta_2 + b_3\beta_3 + b_4\beta_4, \tag{3.64b}$$

$$A_2 = b_1\gamma_1 + b_2\gamma_2 + b_3\gamma_3 + b_4\gamma_4, \tag{3.64c}$$

$$A_3 = b_1\delta_1 + b_2\delta_2 + b_3\delta_3 + b_4\delta_4, \tag{3.64d}$$

$$A_4 = b_1\epsilon_1 + b_2\epsilon_2 + b_3\epsilon_3 + b_4\epsilon_4, \tag{3.64e}$$

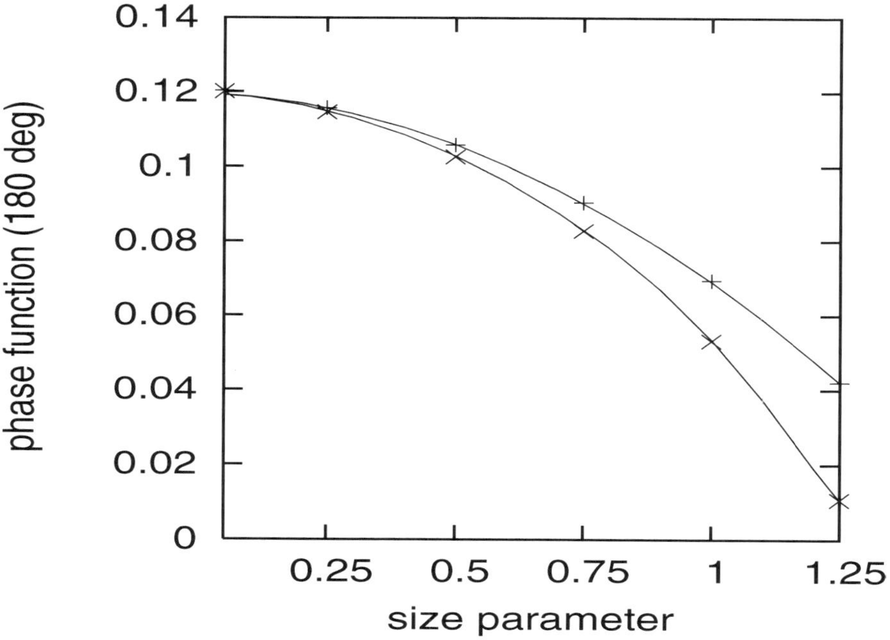

FIGURE 3.2: Comparison of parametrization of $\phi(\pi)$ by (3.63c) with the Mie phase function (solid lines) for two values of relative refractive indices $m = 1.5$ (pluses) and $m = 2.0$ (crosses). This figure was first published in Roy and Sharma (2008).

with

$$b_0 = \frac{15}{8\pi} \cos^2\theta \sin^2\theta, \tag{3.65a}$$

$$b_1 = \frac{1}{4} \cos\theta \left(1 + \cos\theta\right) \left(5\cos^2\theta - 3\right), \tag{3.65b}$$

$$b_2 = \sin^2\theta \left(1 - 5\cos^2\theta\right), \tag{3.65c}$$

$$b_3 = -\frac{1}{4} \cos\theta \left(1 - \cos\theta\right) \left(5\cos^2\theta - 3\right), \tag{3.65d}$$

$$b_4 = b_0/\cos\theta, \tag{3.65e}$$

and

$$\alpha_1 = 0.1187,\ \beta_1 = 0.0045m,\ \gamma_1 = 0.005m^2 - 0.0105m + 0.0411,$$

$$\delta_1 = 0.008m - 0.0048,\ \epsilon_1 = 0.015m - 0.0001 - 0.0151/m, \tag{3.66a}$$

$$\alpha_2 = 0.0597,\ \beta_2 = 0,\ \gamma_2 = -0.0005q,\ \delta_2 = -0.00125q,$$

$$\epsilon_2 = -0.0005\left(1 + 4m^2\right), \tag{3.66b}$$

$$\alpha_3 = 0.12090,\ \beta_3 = -0.007,\ \gamma_3 = -\left(0.0380 + 0.0028p^2\right), \tag{3.66c}$$

$$\delta_3 = -0.0112p^3/m,\ \gamma_3 = -0.00369mp^2, \tag{3.66d}$$

$$\alpha_4 = 0,\ \beta_4 = -0.1235p,\ \gamma_4 = 0.0561m^2 + 0.0217m + 0.0843, \tag{3.66e}$$

$$\delta_4 = \frac{0.0033}{m} - \left(0.126 + 0.06m\right)p,$$

$$\epsilon_4 = \left(0.0033/m^2 + 0.088mp + 0.004p^2\right). \tag{3.66f}$$

3.4.3 Caldas–Semião approximation

Explicit expressions for the small particle phase function have also been obtained by expanding the Mie phase function in powers of the size parameter by many authors in the past (Penndorf 1962, Chu and Churchill 1955). More recently Caldas and Semião (2001a) have revisited the expansion of Mie phase function. For spheres, satisfying the condition $x \leq 1/|m|$ (Rayleigh condition), the phase function was expressed as:

$$i_1(\mu) = |P|^2 x^6 \left[1 + \frac{1}{15}\left(\mathrm{Re}(Q') + \mathrm{Re}(V' + S')\mu\right) x^2 + \frac{4}{3}\mathrm{Im}(P) x^3 + \right.$$

$$\frac{1}{6300}\Big[7|Q'|^2 + 4\mathrm{Re}(R') + 7\left(|V'|^2 + |S'|^2\right)\mu^2 + 14\mathrm{Re}\left[\left(V' + S'\right)Q'^*\right]\mu$$

$$+4\mathrm{Re}\left(W' - T'\right)\mu + 14\mathrm{Re}(S'V')\mu^2 + 20\mathrm{Re}(V')(2\mu^2 - 1) + \mathrm{Re}(U'(5\mu^2 - 1)\Big]x^4$$

$$+\frac{2}{45}\Big[\mathrm{Im}[Q'(P - P^*)] - \mathrm{Im}[(V' + S')P^*]\mu\Big]x^5\Bigg], \qquad (3.67)$$

and

$$i_2(\mu) = |P|^2x^6\Bigg[\mu^2 + \frac{1}{15}\left(\mathrm{Re}(Q')\mu^2 + \mathrm{Re}(V' - S')\mu + 2\mathrm{Re}(S')\mu^3\right)x^2 +$$

$$\frac{4}{3}\mathrm{Im}(P)\mu^2x^3 + \frac{1}{6300}\Big[7|Q'|^2\mu^2 + 4\mathrm{Re}(R')\mu^2 + 7\left(|V'|^2 + |S'|^2\right) +$$

$$28|S'|^2(\mu^4 - \mu^2) + 14\mathrm{Re}\left[\left(V' - S'\right)Q'^*\right]\mu + 4\mathrm{Re}\left(W' + T'\right)\mu + 28\mathrm{Re}(S'Q'^*)\mu^3$$

$$-8\mathrm{Re}(T')\mu^3 + 14\mathrm{Re}(SV'^*)(2\mu^2 - 1) + 20\mathrm{Re}(V')\mu^2 + \mathrm{Re}(U')(15\mu^4 - 11\mu^2)\Big]x^4$$

$$+\frac{2}{45}\Big[\mathrm{Im}[Q'(P - P^*)]\mu^2 - \mathrm{Im}[(V' - S')P^*]\mu - 2\mathrm{Im}(S'P'^*\mu^3\Big]x^5\Bigg]. \qquad (3.68)$$

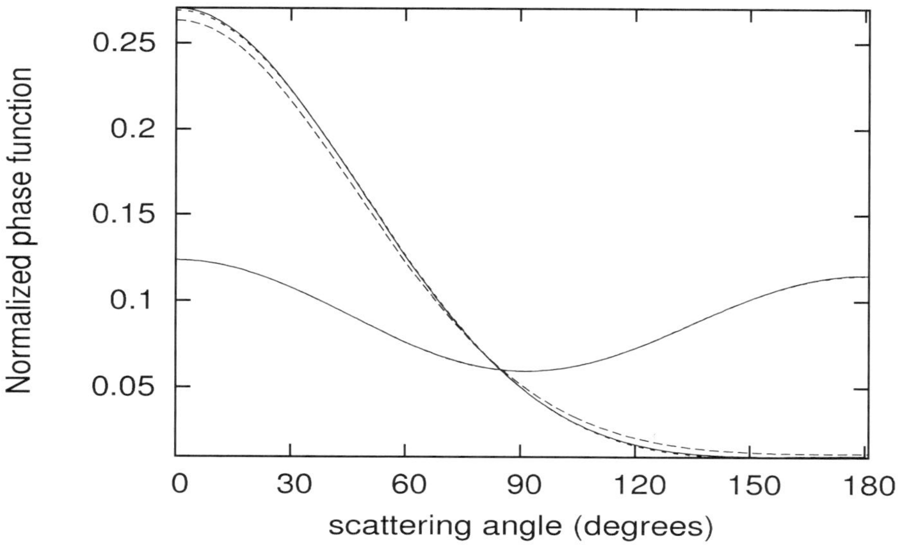

FIGURE 3.3: A comparison of phase functions predicted by $\bar{\phi}_{FPPF}$ and ϕ_{CSPF} with the exact phase function for a sphere of (i) $x = 0.25$ and (ii) $x = 1.25$. In both the cases $m = 2.00$. Solid line is for exact phase function. Dashed line is for ϕ_{CSPF} and small dashed line is prediction of ϕ_{FPPF}. For $x = 0.25$ all the three phase functions overlap with each other.

A comparison of $\bar{\phi}_{FPPF}$ and ϕ_{CSPF} (the phase functions corresponding to 3.67 and 3.68) with Mie phase function has been depicted in the Figure 3.3 for $x = 0.25$ and $x = 1.25$ and $m = 2.0$. It is clear from the figure that for $x = 0.25$ both the approximate phase functions give equally good results (all curves are overlapping). For $x = 1.25$, the differences between the two approximations become observable and $\bar{\phi}_{FPPF}$ can be seen to give a better result.

The dependence of the approximations on x has been shown in Figure 3.4. It displays the root mean square (rms) errors in predictions by $\bar{\phi}_{FPPF}$ and ϕ_{CSPF} against x for $m = 1.5$. The rms error is defined as:

$$\text{rms error}_{CSPF} = \left[\frac{1}{2}\int_0^\pi d\theta \left(\frac{\phi_{CSPF} - \phi_{EX}}{\phi_{EX}}\right)^2\right]^{1/2}, \qquad (3.69)$$

for ϕ_{CSPF}. The same definition holds for the errors in $\bar{\phi}_{FPPF}$ as well.

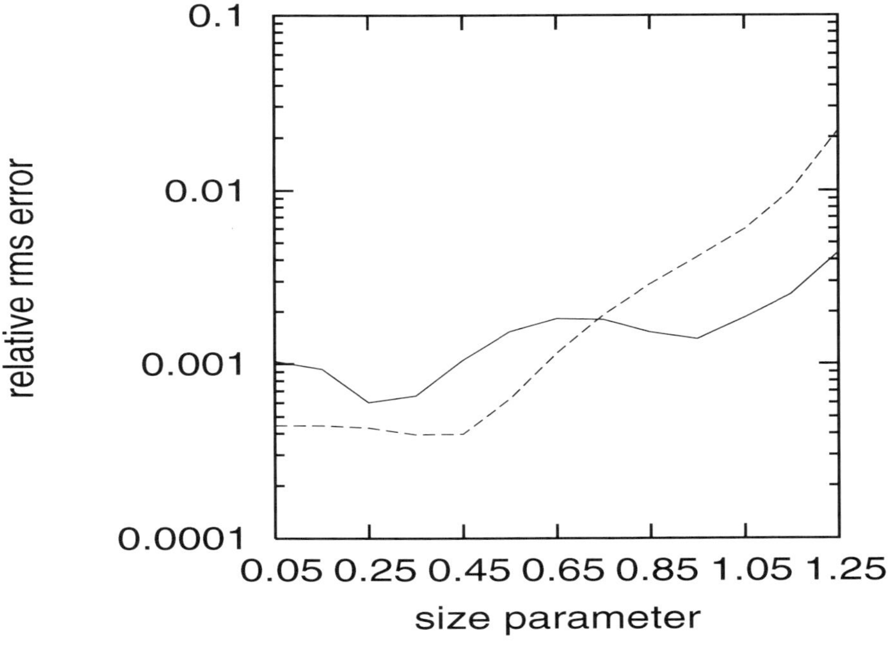

FIGURE 3.4: A comparison of percent rms error in $\bar{\phi}_{FPPF}$ (solid line) and ϕ_{CSPF} (dashed line) as a function of size parameter. The refractive index is 1.50. This figure was first published in Roy and Sharma (2008).

3.5 Rayleigh–Gans approximation

An obstacle is termed soft if it satisfies the condition $|m - 1| \ll 1$. Two approximation methods have been extensively used for predicting the scattering of electromagnetic waves by this class of scatterers. These are (i) the Rayleigh-Gans approximation (RGA) and (ii) the anomalous diffraction approximation (ADA). The latter approximation method is closely related to another approximation, known as the eikonal approximation (EA). A detailed exposition to scattering of light by soft particles, including approximations, can be found in a monograph by Sharma and Somerford (2006).

The RA, discussed in Section 3.2.1, was further developed by Debye (1915), Gans (1925) and many others. Consequently, multiple names have cropped up to refer to this approximation. These include the Rayleigh–Gans approximation (RGA) (van de Hulst 1957, Babenko et al. 2003), the Rayleigh–Debye approximation (Kerker 1969), the Rayleigh–Gans–Debye approximation (Bayvel and Jones 1981), the Rayleigh–Debye-Gans approximation (Stroem 1974), the Rayleigh–Gans–Born approximation (Turner 1976), the Rayleigh–Gans–Rocard approximation (Acquista 1976), etc. It was also adapted in nonrelativistic quantum mechanics where it is known as the Born approximation.

The RGA is expected to work well if,

$$|m - 1| \ll 1, \tag{3.70a}$$

and

$$x|m - 1| \ll 1. \tag{3.70b}$$

It is evident from (3.70b) that the domain of applicability of the RGA allows for $x > 1$ if (3.70a) is complied with. However, objects larger than the wavelength of the wave can no longer be treated as a single dipole. Thus, the scatterer needs to be divided into smaller volume elements. If each of these smaller volume element, acts as an independent dipole, the scattering amplitudes can be written as,

$$S_1(\theta, \varphi)_{RGA} = -ik^3\alpha \; R(\theta, \varphi), \tag{3.71a}$$

and

$$S_2(\theta, \varphi)_{RGA} = -ik^3\alpha \; \cos^2\theta \; R(\theta, \varphi), \tag{3.71b}$$

where

$$\alpha = \frac{(m^2 - 1)\mathcal{V}}{4\pi} \simeq \frac{(m - 1)\mathcal{V}}{2\pi}, \tag{3.72}$$

where $\mathcal{V}$ is the volume of the particle and the shape factor or the form factor, $R(\theta, \varphi)$, is

$$R(\theta, \varphi) = \frac{1}{\mathcal{V}} \int_{\mathcal{V}} e^{i\Delta(\mathbf{r})} d\mathcal{V}, \tag{3.73}$$

with

$$\Delta(\mathbf{r}) = \mathbf{q}.\mathbf{r}, \tag{3.73a}$$

which accounts for the relative phase difference between waves arriving at the detector after being scattered by separate volume elements. This is essentially the same result that was obtained in (2.166) by assuming that the wave inside the particle is not modified by the scatterer. The assumption is plausible because the scatterer is a soft scatterer.

The angular scattering pattern for an unpolarized wave

$$i(\theta, \varphi) = \frac{1}{2} \left[|S_1(\theta, \varphi)|^2 + |S_2(\theta, \varphi)|^2 \right], \tag{3.74}$$

can be expressed as,

$$i(\theta, \varphi)_{RGA} = k^6 \mathcal{V}^2 (1 + \cos^2 \theta) \, \frac{|m - 1|^2}{8\pi^2} \, |R(\theta, \varphi)|^2, \tag{3.75}$$

with the help of (3.71). A number of authors give analytic expressions for $R(\theta, \varphi)$ for scatterers of simple geometrical shapes such as spheres, prolate and oblate spheroids, flat disks, long elliptic cylinders, long circular cylinders, rods, cubes and so on (van de Hulst 1957, Kerker 1969, Bayvel and Jones 1981, Kokhanovsky 2004, Quinten 2011). Analytic expressions for $R(\theta, \varphi)$, have also been obtained for a number of complex particle shapes such as polygonal prisms and hexagonal cylinders, cones, polygonal pyramids, tori, shells and so on (Shapovalov 2014). In addition, aggregates of particles have also been examined and the RGDA results were contrasted with the T-matrix and discrete dipole approximation computations. Appendix I reproduces the shape factors from Shapovalov (2014) for a variety of particle shapes.

The absorption cross section can be obtained by a simple sum of the absorption cross sections of elemental dipoles. This gives,

$$C_{abs}^{RGA} = 2k \int_{\mathcal{V}} n_i(\mathbf{r}) d^3 r, \tag{3.76}$$

which is independent of the shape of the particle. For a homogeneous particle, it is $C_{abs}^{RGA} = 2kn_i\mathcal{V}$.

The scattering cross section defined by (1.64) becomes,

$$C_{sca}^{RGA} = \frac{k^4 \mathcal{V}^2 |m - 1|^2}{2\pi} \mathcal{F}, \tag{3.77}$$

where

$$\mathcal{F} = \frac{1}{2} \int_0^\pi (1 + \cos^2\theta)|R(\theta)|^2 \sin\theta \, d\theta. \tag{3.78}$$

This approach to the RGA can be easily extended to inhomogeneous particles too by assigning different polarizabilities to each elemental dipole.

3.5.1 Homogeneous spheres: Visible and ultraviolet range

For a sphere $\mathcal{V} = (4/3)\pi a^3$. The scattering functions (3.71a) and (3.71b) and the scattered intensity (3.75) then become

$$S_1(\theta)_{RGA} = -ik^3 \frac{m^2-1}{3} R(\theta), \tag{3.79a}$$

$$S_2(\theta)_{RGA} = -ik^3 \frac{m^2-1}{3} \cos^2\theta \; R(\theta), \tag{3.79b}$$

and

$$i(\theta)_{RGA} = \frac{2}{9} x^6 |m-1|^2 (1 + \cos^2\theta)|R(\theta)|^2. \tag{3.79c}$$

For a sphere, the evaluation of the integral in the shape factor is straightforward:

$$R(\theta, \varphi) = \left(\frac{9\pi}{2z^3}\right)^{1/2} J_{3/2}(z) = \frac{3}{z^3}\left(\sin z - z\cos z\right), \tag{3.80}$$

where, as before, $z = 2x\sin(\theta/2)$. By noting that

$$J_{3/2}(z) = \sqrt{\frac{2}{\pi}} \frac{z^{3/2}}{x^2} \sum_{n=0}^{\infty} (2n+1) j_{n+1}^2(x) T_n^1(\cos\theta), \tag{3.80a}$$

and applying the recurrence relation,

$$(2n+3)\mu T_n^1(\mu) = (n+1)T_{n+1}^1(\mu) + (n+2)T_{n-1}^1(\mu), \tag{3.81}$$

equations (3.79) can be converted to a series which is analogous to the series form of Mie scattering phase functions (2.79) and (2.80) and can be easily contrasted with it.

Equation (3.79c), along with (3.80) tells us that the locations of minima in a scattering pattern are governed by the equation $\tan z = z$. Clearly, the positions of minima predicted by the RGA are independent of the particle refractive index. Two modifications to (3.80) appear in the literature. Each introduce an m dependence in the formula. While Shimizu (1983) replaces x by mx, Gordon (1985) argues for substitution of $x = x(1 + m^2 - 2m\cos\theta)^{1/2}$ in place of x in the relation $z = 2x\sin(\theta/2)$. The modified versions leads to considerably better agreement with the Mie theory for positions of extrema in the scattering pattern.

Note that equation (3.80) predicts zero intensity at minima locations. In reality, however, this is not the case. To rectify this shortcoming of the RGA, Gordon (1985) suggested adding an x dependent function to $R(\theta)$:

$$R(\theta) = \frac{3}{z^3}(\sin z - z \sin z) + \gamma(x). \tag{3.82}$$

For $\gamma(x) = x^{-3/2}$, a reasonable agreement with the Mie theory is found.

An important point to be kept in mind, while employing the RGA, is that the extinction theorem,

$$Q_{ext} = (4/x^2)\ \mathrm{Re}S(0), \tag{3.83}$$

is not applicable. For a nonabsorbing particle, the amplitudes $S_1(\theta)$ and $S_2(\theta)$ are purely imaginary. Use of the optical theorem thus gives $Q_{ext}^{RGA} = 0$. Obviously, this is not a correct prediction. This happens because the optical theorem is not applicable to nonunitary amplitudes such as the one obtained in the RGA.

The scattering efficiency factor in the RGA is obtained by using the relation (3.77). This is known to yield:

$$Q_{sca}^{RGA} = |m-1|^2 \tilde{\phi}(x), \tag{3.84}$$

where

$$\tilde{\phi}(x) = \left[\frac{5}{2} + 2x^2 - \frac{\sin 4x}{4x} - \frac{7}{16x^2}(1 - \cos 4x) + \left(\frac{1}{2x^2} - 2\right)(\gamma + \log 4x - Ci(4x))\right], \tag{3.85a}$$

with

$$Ci(x) = -\int_x^{\infty} \frac{\cos t}{t} dt, \tag{3.85b}$$

and $\gamma = 0.577$ is Euler's constant. In the special case of $x \ll 1$, the scattering efficiency factor in the RGA becomes,

$$Q_{sca}^{RGA}(x \ll 1) \approx \frac{32}{27} x^4 |m-1|^2, \tag{3.86}$$

which is same as in the RA for $|m-1| \ll 1$. This is as it should be.

In another special case of $x \gg 1$,

$$Q_{sca}^{RGA} = 2x^2 |m-1|^2, \tag{3.87}$$

which is known as the Jobst approximation (Jobst 1925).

It is easy to see that the absorption efficiency factor is,

$$Q_{abs}^{RGA} = 4x\ \mathrm{Im}\frac{m^2-1}{m^2+1} \quad \simeq \quad \frac{8x}{3}\ \mathrm{Im}(m-1), \tag{3.88}$$

which is the same as in the RA.

The asymmetry parameter takes the following simple form in the RGA (Kokhanovsky and Zege 1997, Irvine 1963):

$$g_{RGA} = \frac{\tilde{\phi}(x)}{H(x)}, \tag{3.89}$$

where

$$H(x) = \frac{4}{x^4}\left[\left(\frac{9}{128} - \frac{5x^2}{64}\right)(4x\sin 4x + \cos 4x) + \frac{x^6}{2} + \frac{11x^4}{8}\right.$$

$$\left. - \frac{31x^2}{64} - \frac{9}{128} + x^2\left(x^2 - \frac{3}{8}\right)(Ci(4x) - \gamma - \ln(4x))\right]. \tag{3.89a}$$

Similar to positions of extrema in the angular scattering pattern, the variation of g against x also does not depend on the refractive index in the RGA.

Numerical comparison of the RGA with exact results has been done by many authors. Earlier works plot the error contour charts (Farone et al. 1963, Heller 1963, Kerker et al. 1963). These charts show error in the (x, m) coordinate plane. It was concluded that the RGA works best for small scattering angles. In particular, it is found that the positions of the first few extrema are reproduced quite faithfully. This observation is suggestive of the possibility that the relation between the extrema position and the size parameter of the sphere can be used to determine the size of the scatterer from the angular scattering pattern. The errors introduced by the use of the RGA in sphere size determination from the position of the maxima in the RGDA have been discussed in detail by Sharma and Somerford (2006).

More recently, Garcia-Lopez et al. (2006) have examined the validity of the RGA for five types of spherical particles over the UV-Visible region (wavelength range $200 \leq \lambda \leq 900\ nm$):

(i) Nonabsorbing soft particles (relative refractive index $m = 1.04$)
(ii) Hemoglobin (relative refractive index $1.01 \leq m \leq 1.2$, $d = 500\ nm,\ 1\ \mu m,\ 5.5\ \mu m$); (relative refractive index $1.01 \leq n_r \leq 1.2,\ 0.001 \leq n_i \leq 0.1,\ d = 500\ nm,\ 1\ \mu m,\ 5.5\ \mu m$)
(iii) Polysterene (refractive index $1.5 \leq n_r \leq 2.2,\ 0.01 \leq n_i \leq 0.85$)
(iv) Silver bromide (relative refractive index $1.1 \leq n_r \leq 2.4$ and $0.0001 \leq n_i \leq 0.85$, $25\ nm \leq d \leq 50\ nm$)
(v) Silver chloride (relative refractive index $1.1 \leq n_r \leq 2.4$ and $0.0001 \leq n_i \leq 0.6$, $25\ nm \leq d \leq 50\ nm$)

It was concluded that the RGA gives adequate approximation for particles of sizes on the order of the magnitude of wavelength. The disagreement becomes significant if strong absorption is present and the particle size is larger than the wavelength of the radiation.

Komar et al. (2013) have derived a modified, semi-analytic expression for the scattering efficiency factor including the effect of strong absorption in the scatterer. This formula constitutes significant improvement over the conventional RGA. For a sphere of $|m| \leq 2$ and $x < 1$, this yields:

$$Q_{sca}^{KKK} = \frac{9}{64} Q_{sca}^{RGA} \left[\frac{1 - e^{-4\tilde{q}a}}{\tilde{q}a} + \frac{1 - (1 + 4\tilde{q}a + 8(\tilde{q}a)^2)e^{-4\tilde{q}a}}{4(\tilde{q}a)^3} + \frac{(1 + 4\tilde{q}a)e^{-4\tilde{q}a} - 1}{2(\tilde{q}a)^2} \right]^2, \tag{3.90}$$

where $\tilde{q} = k\sqrt{0.05 n_i |m - 1|}$. The superscript KKK stands for Komar, Kocifaj and Kohut. As $n_i \to 0$ or $a \to 0$, the correction term approaches 1. On the other hand, if $x \to \infty$, the $Q_{ext}^{RGA} \propto x^2$ (see eq. (3.87)) and the correction term is proportional to $1/x^2$. Therefore, it appears plausible that (3.90) can be valid even for moderately larger values of x. Numerical comparisons indeed confirm that the Q_{sca}^{KKK} simulates the scattering efficiency factor much better than the conventional RGA.

A different approach has been conceived by Garcia-Lopez et al. (2008). The idea is to use an approximate Mie solution to obtain the field inside the particle, which then generates the elementary dipoles. This takes into account the effect of the dipole field alterations of the incoming field. This is in contrast to the RGA, where it is assumed that all the dipoles are a consequence of the same field, namely, the incident field. The field inside the particle is approximated to be,

$$E(\mathbf{r}) \approx \frac{3}{2} c_1 M_{011} - \frac{3}{2} i d_1 N_{e11} - \frac{5}{6} i d_2 N_{e12}. \tag{3.91}$$

The terms neglected are on the order of $(a/\lambda)^2$. Using this as the field responsible for generating dipoles, it is possible to write,

$$\begin{pmatrix} E_{\parallel}^s \\ E_{\perp}^s \end{pmatrix} = -\frac{3k^2\mathcal{V}}{4\pi r} \frac{m^2 - 1}{m^2 + 2} \begin{pmatrix} -f_1 \cos\theta + f_2 \frac{\sin\theta}{\cos\varphi} & 0 \\ 0 & f_1 \end{pmatrix} \begin{pmatrix} E_{\parallel}^i \\ E_{\perp}^i \end{pmatrix}. \tag{3.92}$$

where

$$f_1 = \frac{2\pi}{\mathcal{V}} \left(\frac{iAae^{iAa} - e^{iAa} + iaAe^{-iaA} + e^{-iaA}}{i^3 A^3} \right), \tag{3.93a}$$

$$f_2 = \frac{2\pi}{\mathcal{V}} \left[\left(\frac{iBae^{iBa} - e^{iBa} + iaBe^{-iaB} + e^{-iaB}}{i^3 B^3} \right) - \right.$$

$$\left(\frac{iCae^{iCa} - e^{iCa} + iaCe^{-iaC} + e^{-iaC}}{i^3C^3}\right)\Bigg], \tag{3.93b}$$

$$A = \sqrt{k^2 + \frac{k_1^2(d_2 + c_1)^2}{4d_1^2} - \frac{kk_1(d_2 + c_1)}{d_1}\cos\theta}, \tag{3.93c}$$

$$B = \sqrt{k^2 + \frac{k_1^2(d_2 - c_1)^2}{4} - \frac{kk_1(d_2 - c_1)}{d_1}\cos\theta\cos\varphi}, \tag{3.93d}$$

$$C = k. \tag{3.93e}$$

In the above equations, k is the wave number in the host medium and k_1 is the wave number in the sphere. This approximation has demonstrated vastly improved accuracy and applicability for a broader range of optical parameters.

An interpolation formula for a dielectric sphere, based on Rayleigh-Gans theory, was given by Walstra (1964). If, for a given x, Q_{ext} is known for $m = a$ and $m = c$, Q^{ext} can be computed for the intermediate value of $m = b$ by

$$Q_{ext}(b) = \frac{(b-1)^2}{c-a}\left[\frac{c-b}{(a-1)^2}Q_{ext}(a) + \frac{b-a}{(c-1)^2}Q_{ext}(c)\right]. \tag{3.94}$$

The results from this approximation yield less than 1 percent error for values of x up to at least 8. The m range considered was $m = 1.025(0.05)1.275$.

For a small range of ρ values, $1.5 \leq \rho \leq 2.5$, a simple formula

$$Q_{ext}^{WAA} = (1.26m - 0.04)\rho - 2.558(m-1)^{1.273} - 0.843, \tag{3.95}$$

gives less than 1 percent error (Walstra 1964).

3.5.2 Homogeneous spheres: X-ray energies

At X-ray energies, it is possible to express the refractive index in terms of the density of the material. The condition (3.70b) can be re-expressed as (Windt et al. 2000),

$$\frac{a(\mu m)}{E(keV}\left(\frac{\rho}{3(g\ cm^{-3})}\right) \ll 0.316, \tag{3.96}$$

where a is the radius in micron, E is the X-ray energy in keV and ρ is the mass density of the material in gm/cm^3. A detailed investigation of the validity of the RGA for simulation of small angle scattering by particles of interest in interstellar dust studies concluded that the RGA is a good approximation for small angle X-ray scattering if $E \geq 1\ keV$ (Smith and Dwek 1998).

In the limit of small angles, it is possible to approximate,

$$\left(\frac{J_{3/2}(x)}{x^{3/2}}\right)^2 \approx \frac{2\pi}{9}\exp\left(\frac{-z^2}{2\sigma^2}\right). \tag{3.97}$$

The scattered intensity (3.79c) can therefore be written as (see, for example, Mauche and Gorenstein 1986),

$$i(\theta)_{RGA} = \frac{2}{9}a^2x^4|m-1|^2\exp\left(-\frac{z^2}{2\sigma^2}\right)(1+\cos^2\theta), \tag{3.98}$$

where

$$\sigma = \frac{10.4}{E(keV)\ a(\mu m)}. \tag{3.99}$$

The Gaussian approximation to the Bessel function has been found to be adequate only if $E > 2\ keV$ (Smith and Dwek 1998).

Employing a simplified form of the Ramsauer approximation (Ramsauer 1921, Louedec and Urban 2012), Sharma (2015) suggested a modification of the RGA scattered intensity. The modified formula reads as:

$$i(\theta)_{MRGA} = i(\theta)_{RGA} - 150\lambda a^2x^4|m-1|^2(1+\cos^2\theta) \times\left[\left(\frac{j_1(z)}{z}\right)^2 - \frac{1}{2}\left(\frac{J_1(x\sin\theta)}{x\sin\theta}\right)^2\right]. \tag{3.100}$$

Numerical evaluation of this approximation has been done in the size range $0.0025\ \mu m \leq a \leq 0.25\ \mu m$. This range corresponds to particles of interest in the interstellar dust. It is found that the MRGA indeed constitutes an improvement over the traditional RGA. It extends the angular range of the approximation as well as the range of X-ray energies over which the approximation is valid. A typical comparison is shown in Figure 3.5 for near-forward scattering. In this figure $x \approx 3$. It can be seen that while the RGA differs significantly from the Mie theory, the MRGA shows good agreement with the Mie results.

3.5.3 Nonspherical particles

The accuracy of the RGA and some of its modified variants has been examined for nonspherical particles too. Barber and Wang (1978) examined it for spheroids and showed that for $m = 1.05$ and $2x|m-1| \leq 1.0$, the errors up to 20% are generated depending on the orientation of the spheroid. The worst case occurs where the large axis is parallel to the incident electric vector. The errors in the RGA for ellipsoidal particles have been examined by Latimer and coworkers for $m < 1.5$ (Latimer 1975, 1980; Latimer and Barber 1978).

The quasi-static approximation (QSA) is another attempt to generalize the RGA to nonspherical shapes (Burberg 1956, Shatilov 1960, Voshchinnikov and Farafonov 2000, Farafonov et al. 2001, Posselt et al. 2002). In this approximation the field inside the scatterer is taken to be the incident field, while polarizability is considered to be that in the RA. The condition for applicability of the QSA is (see, for example, Posselt et al. 2002),

$$|m-1|x_v \ll 1,$$

where, in the case of a spheroid, $x_v = 2\pi a_v/\lambda$ (a_v is the radius of a sphere whose volume is equal to that of the spheroid). It has been shown that the QSA is preferable to the RGA for optically soft particles. In comparison to the RGA and the RA, its range of applicability is nearly always greater if $a/b \geq 3$, a and b being semi-major and semi-minor axes, respectively.

For a cylinder of length l and radius a, the scattering functions for a tilted cylinder incidence are (van de Hulst 1957)

$$\begin{pmatrix} S_1 \\ S_2 \end{pmatrix} = \frac{-ikl(m^2-1)x^2}{4} \begin{pmatrix} 1 \\ \cos\theta \end{pmatrix} R(\theta, \varphi, \beta), \tag{3.101}$$

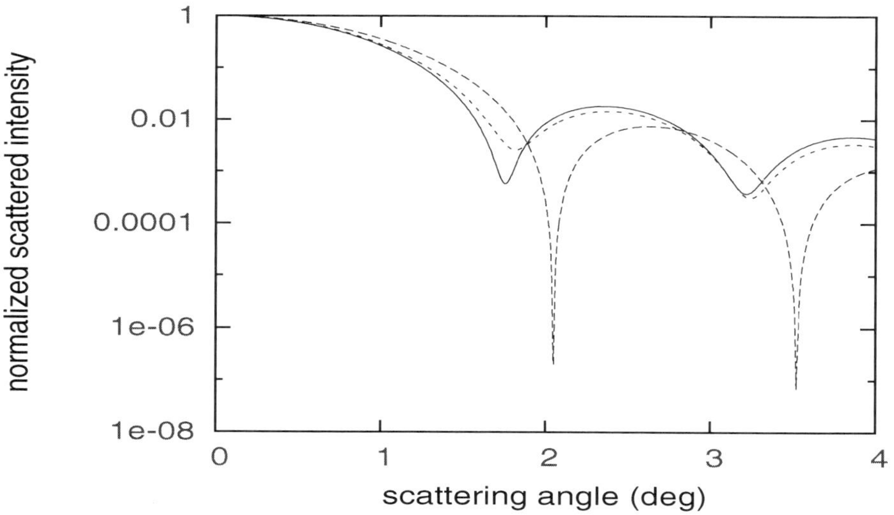

FIGURE 3.5: A comparison of MRGA with RGA and the Mie phase function. Here, $a = 0.2$ μm, $\lambda = 0.00999871$ μm. The solid line is the Mie phase function, dashed line is RGA and the dash-dot line is the MRGA.

where, β is the angle between the positive z axis and the incident wave. The shape factor $R(\theta, \varphi, \beta)$ is then given by (Bohren and Huffman 1983),

$$R(\theta, \varphi, \beta) = \frac{2\sin(x\mathcal{L}\mathcal{A})}{x\mathcal{L}\mathcal{A}} \frac{J_1(xM)}{xM}, \tag{3.102}$$

where $\mathcal{L} = l/a$, $A = \cos\beta + \sin\theta\cos\varphi$, $M = \sqrt{B^2 + C^2}$, $B = \sin\theta\sin\varphi$, and $C = \cos\theta - \sin\beta$. The TMWS and the TEWS functions for an infinitely long cylinder at perpendicular incidence can be obtained from (3.101) by using the relation (van de Hulst 1957),

$$\begin{pmatrix} T_{TMWS} \\ T_{TEWS} \end{pmatrix} = \frac{\pi}{kl} \begin{pmatrix} S_1 \\ S_2 \end{pmatrix}. \tag{3.103}$$

The scattering functions for scattering by an infinitely long cylinder in the RGA are thus:

$$T(\theta)_{TMWS}^{RGA} = \frac{-i\pi(m-1)x^2 J_1(\tilde{z})}{\tilde{z}}, \tag{3.104a}$$

and

$$T(\theta)_{TEWS}^{RGA} = \cos\theta\; T(\theta)_{TMWS}^{RGA}. \tag{3.104b}$$

In (3.104) $\tilde{z} = x\sin\theta$. For perpendicular incidence, $\beta = \pi/2$, $R(\theta, \varphi)$ reduces to $2J_1(\tilde{z})/\tilde{z}$. It is evident that the positions of minima in the scattering patterns are governed by the minima of $J_1(\tilde{z})/\tilde{z}$. Modifications analogous to those described by Shimizu (1983) and Gordon (1985) for a sphere have been studied for an infinitely long cylinder by Sharma and Somerford (1988). These are found to reproduce the positions of extrema more fittingly.

Another empirical modification of the RGA has been proposed by Zhao et al. (2006) for a small particle limit:

$$i(\theta)_{small} = \frac{k^2 C_{sca}}{4\pi^2} a_0 \left[t|b_1|^2 + (1-t)\left(|b_2| + \gamma\right)^2\right]\left(1 + \cos^2\theta\right), \tag{3.105}$$

where b_1 is nothing but the form factor given by (3.73),

$$b_2 = \frac{1}{\mathcal{V}} \int e^{i(m\mathbf{k}_i - \mathbf{k}_s).\mathbf{r}'} d^3 r', \tag{3.106a}$$

$$t = \exp\left(-c_1 x_{vp}^3\right), \tag{3.106b}$$

$$\gamma = \frac{x_{vp}^{4.5}}{\left(200 + x_{vp}^6\right)\left(1 + m^2 - 2m\cos\theta\right)^{3/4}}, \tag{3.106c}$$

$$c_1 = 5\ \mathrm{Re}\left[\frac{(m-1)}{8}\right], \tag{3.106d}$$

and

$$x_{vp} = \frac{3k\mathcal{V}}{4P}, \tag{3.106e}$$

is the equivalent size parameter sphere with P as the projected area.

3.6 The eikonal approximation

It was shown in (1.78) that the general expression for the scattering function can be written as,

$$S(\theta) = \frac{ik^3}{4\pi} \int d\mathbf{r}\ e^{-i\mathbf{k_s}.\mathbf{r}} \Big[1 - m^2(\mathbf{r})\Big] \Psi(\mathbf{r}), \tag{3.107}$$

where $\Psi(\mathbf{r})$ is the scalar electromagnetic field inside the scatterer. In the EA, $\Psi(\mathbf{r})$ is approximated as,

$$\Psi(\mathbf{r}) \approx \exp\left[i\mathbf{k}_i.\mathbf{r} + \frac{ik}{2}\int_{-\infty}^{z} (m^2(\mathbf{b}, z') - 1)dz'\right]. \tag{3.108}$$

The scattering picture in the EA is shown in Figure 3.6. A ray enters or leaves the particle undeviated from its initial trajectory with almost no change in the amplitude. The assumption of no change in the amplitude is good if,

$$|m - 1| \ll 1. \tag{3.108a}$$

Straight line propagation of the wave inside the particle demands that the refractive index of the particle vary slowly in a wavelength. Mathematically, this condition may be expressed as,

$$ka \gg 1. \tag{3.108b}$$

The straight line propagation inside the particle ensures that the phase shift suffered by the ray is as given in (3.108). If the refractive index has an imaginary part, the condition (3.108a) is equivalent to two conditions:

$$|n_r(\mathbf{r}) - 1| \ll 1, \tag{3.109a}$$

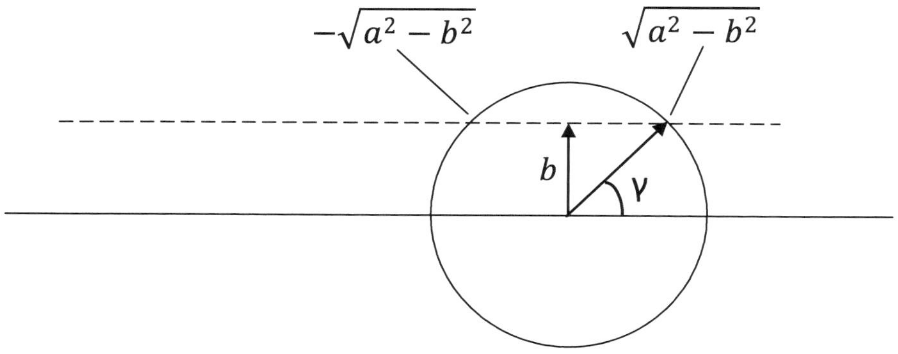

FIGURE 3.6: Ray propagation in the eikonal approximation. The impact parameter is denoted by b.

and

$$n_i(\mathbf{r}) \ll 1. \tag{3.109b}$$

The maximum value of the refractive index within the particle may be taken for the purpose of inequalities.

Substitution of (3.108) in (3.107) gives,

$$S(\theta) = \frac{ik^3}{4\pi} \int d\mathbf{b}dz \; e^{i\mathbf{q}.\mathbf{r}} \left[1 - m^2(\mathbf{r})\right] \exp\left[\frac{ik}{2} \int_{-\infty}^{z} (m^2(\mathbf{b}, z') - 1) \; dz'\right]. \tag{3.110}$$

For a small angle scattering, q_z may be approximated as,

$$q_z = 2k \sin(\theta/2) \approx 0. \tag{3.111}$$

Then the integration over z can be performed to yield,

$$S(\theta)_{EA} = -k^2 \int d\mathbf{b} e^{i\mathbf{q}.\mathbf{b}} \left[e^{i\chi(\mathbf{b})_{EA}} - 1\right], \tag{3.112}$$

where

$$\chi(\mathbf{b})_{EA} = (k/2) \int_{-\infty}^{\infty} \left[m^2(\mathbf{r}) - 1\right] dz, \tag{3.112a}$$

and $\mathbf{q}.\mathbf{b} = 2kb \sin(\theta/2) \cos\varphi$. The two-dimensional integration is over the projected area of the particle and the phase shift $\chi_{EA}(\mathbf{b})$ depends on the size, shape and refractive index profile of the scatterer.

The angular range, resulting from the small angle approximation translates to

$$\theta \ll \frac{1}{\sqrt{x}} \qquad \text{for} \qquad x|m^2 - 1| < 1, \tag{3.113a}$$

and

$$\theta \ll |m^2 - 1|^{1/2} \qquad \text{for} \qquad x|m^2 - 1| > 1. \tag{3.113b}$$

More details can be found in Sharma and Somerford (2006). In summary, the EA is a $|m - 1| \to 1$ approximation valid for small angle scattering.

The approximate form (3.107) can also be obtained by linearizing the propagator in (1.75). For this, one may write

$$G(\mathbf{r} - \mathbf{r}') = -\frac{1}{(2\pi)^3} \int d\mathbf{p} \frac{e^{i\mathbf{p}.(\mathbf{r}-\mathbf{r}')}}{p^2 - k^2 - i\epsilon}, \tag{3.114}$$

where $\mathbf{p}$ is an integration variable. Defining a new variable $\mathbf{Q} = \mathbf{p} - \mathbf{k}_i$, the propagator can be expressed as,

$$G(\mathbf{r} - \mathbf{r}') = -\frac{e^{i\mathbf{k}_i.(\mathbf{r}-\mathbf{r}')}}{(2\pi)^3} \int d\mathbf{Q} \frac{e^{i\mathbf{Q}.(\mathbf{r}-\mathbf{r}'}}{2\mathbf{k}_i.\mathbf{Q} + \mathbf{Q}^2 - i\epsilon}. \tag{3.115}$$

Because of the assumed near-forward propagation, $|\mathbf{Q}| \ll k$. Thus, neglect of the Q^2 term in the denominator of the propagator gives,

$$G(\mathbf{r}-\mathbf{r}')_{EA} = \frac{-i}{2k} e^{ik(z-z')} \delta(\mathbf{b}-\mathbf{b}')\Theta(z-z'), \tag{3.116}$$

where Θ is the step function and δ is the Dirac delta function. The linearized propagator clearly exhibits a straight line propagation in the $\mathbf{k}_i$ direction. Use of this propagator in the integral equation (1.75) then leads to (3.108).

3.6.1 Homogeneous spheres

For a homogeneous sphere $m(\mathbf{r}) = m$. Thus, the integration over the azimuthal angle in (3.112) is straightforward. The scattering function may then be set as

$$S(\theta)_{EA} = -k^2 \int_0^a b\, db\, J_0\Big(2kb\sin(\theta/2)\Big)\Big[\exp\Big(i\chi(b)_{EA}\Big) - 1\Big], \tag{3.117}$$

where

$$\chi(b)_{EA} = x(m^2-1)\sqrt{[1-(b^2/a^2)]} = \tilde{\rho}_{EA}\sqrt{[1-(b^2/a^2)]}, \tag{3.117a}$$

with

$$\tilde{\rho}_{EA} = x(n_r^2 - n_i^2 - 1) + 2ixn_rn_i \approx x(n^2-1) + 2ixn_rn_i \equiv \rho_{EA} + i\kappa_{EA}. \tag{3.117b}$$

It can be seen from Figure 3.6 that the distance travelled by the ray in the medium at an impact parameter b is $2\sqrt{a^2-b^2}$. Thus, ρ_{EA} is nothing but the phase shift suffered by the central ray in passing through a diameter.

Substituting $b = a\sin\gamma$, an often used alternative form of (3.112) is,

$$S(\theta)_{EA} = x^2 \int_0^{\pi/2} J_0(z)\sin\gamma\left[1 - e^{\tilde{\omega}_{EA}\cos\gamma}\right]\cos\gamma\sin\gamma d\gamma, \tag{3.118}$$

where $\tilde{\omega}_{EA} = i\tilde{\rho}_{EA}$. For forward scattering, (3.118) yields a closed form analytic expression:

$$S(0)_{EA} = x^2\left(\frac{1}{2} - \frac{e^{\tilde{\omega}_{EA}}}{\tilde{\omega}_{EA}} + \frac{e^{\tilde{\omega}_{EA}} - 1}{\tilde{\omega}_{EA}^2}\right). \tag{3.119}$$

Employing the extinction theorem,

$$Q_{ext} = \frac{4}{x^2}\mathrm{Re}S(0), \tag{3.120}$$

the extinction efficiency factor can be cast as:

$$Q_{ext}^{EA} = \Bigg[2 - \frac{4\cos\beta_{EA}}{\rho_{EA}}\sin(\rho_{EA} - \beta_{EA})\exp(-\rho_{EA}\tan\beta_{EA})$$

$$+4\left(\frac{\cos\beta_{EA}}{\rho_{EA}}\right)^2\cos(2\beta_{EA})$$

$$-4\left(\frac{\cos\beta_{EA}}{\rho_{EA}}\right)^2\cos(\rho_{EA}-2\beta_{EA})\exp(-\rho_{EA}\tan\beta_{EA})\Bigg], \quad (3.120a)$$

for an absorbing sphere, with

$$\tan\beta_{EA}=\frac{2n_r n_i x}{\rho_{EA}}, \quad (3.120b)$$

and

$$Q_{ext}^{EA}=\left[2-\frac{4}{\rho_{EA}}\sin\rho_{EA}-+\frac{4}{\rho_{EA}^2}(1-\cos\rho_{EA})\right], \quad (3.120c)$$

for a nonabsorbing sphere ($\beta_{EA}=0$).

For nonforward scattering, closed analytic forms for $S(\theta)$ can be obtained in two special cases:

(i) If $n_i\to\infty$:

$$S(\theta)_{EA}=x^2\frac{J_1(z)}{z}, \quad (3.121a)$$

where $z=2x\sin(\theta/2)$. It may be recognized that this is nothing but the scattering function for a sphere in the Fraunhofer diffraction approximation.

(ii) If $x|m^2-1|\ll 1$, the exponential in (3.118) may be approximated as,

$$\exp[\tilde{\omega}_{EA}\cos\gamma]\approx 1+\tilde{\omega}_{EA}\cos\gamma. \quad (3.121b)$$

Consequently, the scattering function may be written as,

$$S(\theta)_{EA}=i(m^2-1)x\left[\cos z-\frac{\sin z}{z}\right]/2\sin^2(\theta/2), \quad (3.121c)$$

which may be recognized as the scattering function for a homogeneous sphere in the RGA for the near forward scattering. This implies that despite the initial premise $x\gg 1$, required for its validity, the EA may prove to be a good approximation for any x value if the condition $\rho_{EA}\ll 1$ is met.

If the refractive index of the particle is real, a closed form analytic expression can be obtained for Im $S(\theta)_{EA}$, for arbitrary scattering angles:

$$\text{Im } S(\theta)_{EA}=-\rho_{EA}\frac{x^2}{\hat{y}^2}\sqrt{\frac{\pi\hat{y}}{2}}J_{3/2}(\hat{y}), \quad (3.122)$$

where

$$\hat{y}^2=\rho_{EA}^2+z^2, \quad (3.122a)$$

and $z = 2x\sin(\theta/2)$. The real part of the scattering function can only be expressed as a series of known functions. For $\rho_{EA} < 1$ it yields,

$$\text{Re } S(\theta)_{EA} = x^2 \left[\left(\frac{\rho_{EA}}{z}\right)^2 J_2(z) - \frac{\rho_{EA}^4}{1.3}\frac{1}{z^3}J_3(z) + \frac{\rho_{EA}^6}{1.3.5}\frac{1}{z^5}J_4(z) + ... \right], \tag{3.123a}$$

and

$$\text{Re } S(\theta)_{EA} = x^2 \left[\frac{1}{z}J_1(z) + \frac{\rho_{EA}}{\hat{y}^{3/2}}\sqrt{\pi/2}N_{3/2}(\hat{y}) + \frac{1}{\rho_{EA}^2}J_0(z) + \frac{1.3}{\rho_{EA}^4}zJ_1(z) + \frac{1.3.5}{\rho_{EA}^6}z^2J_2(z) + ... \right], \tag{3.123b}$$

for $\rho_{EA} > z$ (see, for example, van de Hulst 1957).

The absorption efficiency factor can be obtained in a straight forward manner from the relationship,

$$Q_{abs}^{EA} = \frac{1}{\pi a^2} \int d\mathbf{b} \left[1 - \exp\left(-\text{Im}(\chi_{EA})\right)\right], \tag{3.124}$$

to yield,

$$Q_{abs}^{EA} = 2\mathcal{K}\left(-2\kappa_{EA}\right) = 1 + \frac{e^{-2\kappa_{EA}}}{\kappa_{EA}} + \frac{(e^{-2\kappa_{EA}} - 1)}{\kappa_{EA}^2}. \tag{3.124a}$$

When $\kappa_{EA} \gg 1$, all rays incident to the sphere are absorbed and Q_{abs} is close to 1.

3.6.2 Corrections to the eikonal approximation

Corrections to the eikonal approximation, obtained in the context of potential scattering, have been translated to electromagnetic wave scattering using (1.73). Based on the corrections given by Wallace (1973) in potential scattering, Sharma et al. (1982) suggested the following first-order correction to the EA:

$$S(\theta)_{FC} = x^2 \int J_0(z\sin\gamma) \left[1 - e^{ix(m^2-1)\cos\gamma}\left(1 + i\tau_1\right)\right] \sin\gamma\cos\gamma d\gamma, \tag{3.125}$$

where

$$\tau_1 = \frac{x|m^2-1|^2\cos\gamma}{4}\left(1 - \tan^2\gamma\right). \tag{3.125a}$$

The abbreviation *FC* refers to the first-order correction. The validity of (3.125) is governed by the requirement:

$$\frac{x|m^2-1|^2}{4} \ll 1. \tag{3.126}$$

For forward scattering, (3.125) yields a closed form analytic expressions

$$S(0)_{FC} = m^2 S(0)_{EA} + \frac{x^2}{4}(m^2-1)\left[e^{-ix(m^2-1)} - 1\right]. \tag{3.127}$$

The next higher-order correction is

$$S(0)_{SC} = C(m)S(0)_{ADA}, \tag{3.127a}$$

with

$$C(m) = 1 + 0.25(m+1)^2(m-1)(2-m^2). \tag{3.127b}$$

$S(0)_{ADA}$ is a forward scattering function in the anomalous diffraction approximation (ADA)(see, Section 3.7). The correction (3.127) and (3.127a) clearly preserve the simplicity of the unmodified approximation.

The angular range of the validity of the EA is expected to increase if the small angle approximation (3.111) is not made. The resulting formalism is tagged as the eikonal picture (EP). The z integration can still be performed and the scattering function in the EP can be cast as (Perrin and Chiappetta 1985),

$$S(\theta)_{EP} = -k^3(m^2-1)\int_0^a b\, db\, J_0(kb\sin\theta) \times \left[\frac{e^{i[q_z + k(m^2-1)]\sqrt{a^2-b^2}} - e^{-iq_z\sqrt{a^2-b^2}}}{2q_z + k(m^2-1)}\right] \tag{3.128}$$

where $q_z = 2k\sin^2(\theta/2)$. It may appear from (3.128) that when $\mathrm{Im}|m^2-1|$ is negligible and $4\sin^2(\theta/2) + \mathrm{Re}|m^2-1| = 0$, resonances are produced. But it can be easily checked that these resonances are spurious.

For $\theta = 0$, $\exp(2ikz\sin(\theta/2) = 1$, and the EP amplitude is identical to the EA amplitude. If $\mathrm{Im}(mx) \geq 1$, the first term in the square bracket in (3.128) may be neglected. For the near-forward direction, this leads to diffractive scattering. For backward scattering, this leads to a correct geometrical optics result (Bourrley et al. 1996) if $(m^2-1)/2q_z + k(m^2-1) = r_\perp(\theta)$, $r_\perp$ as the Fresnel coefficient for the perpendicular component of the electric field. A variant similar to this was suggested by Berlad (1971) in the context of scattering of spin 1/2 particles.

As is clear from (3.127a), the EA is closely related to the ADA. The ADA will be considered in detail in the next section. In addition, the EA is also connected closely to another approximation known as the Rytov approximation (1937). The Rytov approximation assumes the solution of (1.72) in the form,

$$\Psi(\mathbf{r}, \xi) = e^{R(\mathbf{r},\xi)}, \tag{3.129}$$

where

$$R(\mathbf{r},\xi) = \sum_{n=0}^{\infty} \xi^n R_n(\mathbf{r}), \tag{3.129a}$$

and ξ is a coupling constant. It has been found that the lowest-order term in the expansion leads to a scattering amplitude which is the same as the EA. The first-order corrections however differ (Car et al. 1977). The Rytov approximation has been contrasted extensively with the RGA, and the Born approximation, as well (see, for example, Marks 2006 and references therein).

3.6.3 Generalized eikonal approximation

The generalized eikonal approximation (GEA) employs a linearized propagator with two adjustable parameters. The two parameters, α_0 and δ, are determined by requiring that the GEA correctly leads to the ray optics phase shift and the edge effects. The scattering function in the GEA takes the form (Chen 1989),

$$S(\theta)_{GEA} = \frac{ik}{4\pi}(1-\delta)S_B - (m^2-1)\alpha_0\delta^2\, k^2 I, \tag{3.130}$$

where

$$I = a^2 \int_0^{\pi/2} J_0(z\sin\gamma)\left[e^{i\tilde{\rho}_{GEA}\cos\gamma} - 1\right]\cos\gamma\sin\gamma\; d\gamma, \tag{3.130a}$$

$$S_B(\theta) = -4x^2(m^2-1)a\int_0^{\pi/2} \sin^2\gamma\cos\gamma J_0(z\sin\gamma)\; d\gamma, \tag{3.130b}$$

$$\delta = (m+1)/2\alpha_0, \qquad \tilde{\rho}_{GEA} = x(m^2-1)/\alpha_0\delta, \tag{3.130c}$$

and

$$\alpha_0 = \frac{m+1}{2} - \frac{3i}{8}\left[\frac{1}{x} - \frac{2}{\tilde{\rho}_{GEA}}\left(\frac{\alpha_1}{x^{2/3}} - \frac{\alpha_2}{x^{4/3}}\right)\right], \tag{3.130d}$$

with $\alpha_1 = 2+2.4i, \alpha_2 = 2+6i$. For forward scattering, (3.130) can be evaluated analytically to yield:

$$S(0)_{GEA} = -\frac{i}{3}\alpha_0\delta x^2(1-\delta)+\frac{1}{2}\alpha_0\delta^2x^2\left[1+\frac{2i}{\tilde{\rho}_{GEA}}e^{i\tilde{\rho}_{GEA}}+\frac{2}{\tilde{\rho}^2_{GEA}}\left(1-e^{i\tilde{\rho}_{GEA}}\right)\right]. \tag{3.131}$$

Equation (3.131) reduces to the EA for $\delta = 1$ and $\alpha_0 = 1$.

Numerical evaluation of the GEA for the extinction efficiency factor and scattered intensity show improvement over the EA results (Chen 1989). The accuracy of the GEA for Q_{ext} at $\lambda = 0.73\ \mu m$ has been rendered by Wang et al. (2012). The errors are less than 6% for $d > 1\ \mu m$. The GEA and the Mie results coincide when the oscillations in the extinction curve are small. But there are large errors in the small diameter region. It has been suggested that one could use Mie theory for $d < 1\ \mu m$ and the GEA for $d > 1\ \mu m$ (Wang et al. 2012). The method has been called the combined GEA (CGEA).

It has been argued that a more logical way to compute the phase shift along a straight line propagation is to do so along a direction given by $(\mathbf{k}_i + \mathbf{k}_s)/2$ and not along $\mathbf{k}_i$ as is normally done in the conventional EA. This is equivalent to conjecturing that the ray deviates by an angle $\theta/2$ at each boundary. The amended phase shift in the modified GEA (MGEA) is (Chen 1993):

$$\tilde{\rho}_{MGEA} = 2x(m - \cos\theta/2), \tag{3.132}$$

and,

$$\delta = (m^2 - 1)/\left[2\alpha_0[m - \cos(\theta/2)]\right], \tag{3.133}$$

is the new relation between α_0 and δ. The scattering function for a large sphere can be evaluated analytically in the MGEA, leading to (Chen 1993),

$$S(\theta)^{LS}_{MGEA} = (m^2 - 1)x^3\left[-i(1-\delta)\frac{j_1(z)}{z} + \frac{\delta J_1(z)}{z\tilde{\rho}_{MGEA}} + \frac{\delta}{\hat{y}^2}\left(ie^{i\hat{y}} + \frac{1 - e^{i\hat{y}}}{\hat{y}}\right)\right], \tag{3.134}$$

where α_0 is redefined as

$$\alpha_0 = \frac{m+1}{2} - \frac{3i}{8x}, \tag{3.134a}$$

and $\hat{y} = [z^2 + \tilde{\rho}_{MGEA}]^{1/2}$. The superscript LS refers to large scatterers.

For small scattering angles and to the lowest order in $1/x$, (3.134) simplifies to,

$$S(\theta)^{LS}_{MGEA} = (m^2 - 1)x^3\delta\left[\frac{J_1(z)}{\tilde{\rho}_{MGEA}} + \frac{3j_1(z)}{4x(m+1)z} + i\frac{e^{i\hat{y}}}{\hat{y}^2} + \frac{1 - e^{i\hat{y}}}{\hat{y}^3}\right]. \tag{3.135}$$

From (3.135), it is straightforward to derive the following relationship for the location of the first minimum in the scattering pattern of a nonabsorbing sphere (Chen 1994):

$$2a\sin\theta_1 = \left[1.220 - \frac{(m-1)/(m+1) + 15.87\sin y_1/y_1}{\pi(1.667 - 12.43\sin y_1/y_1)}\right], \tag{3.136a}$$

where

$$y_1 = 2x\sqrt{m^2 + 1 - 2m\sqrt{1 - (1.916/x)^2}}. \tag{3.136b}$$

In a similar way, the position of the second minimum has also been related to m and x. The predictions of these relations are found to be in excellent agreement with the exact results for $x > 50$. The refractive index values in this investigation were $m = 1.2$ and $m = 1.33$. Since the EA and its variants are small angle approximations, only the first few minima positions are expected to yield reliable predictions.

Formulas analogous to (3.136a) have been obtained by Sharma et al. (1997a) for a sphere and for an infinitely long cylinder. For a sphere it reads,

$$x \sin\theta_1 = 3.832 - \frac{\rho \sin(y_1)}{0.105 y_1^2 - 0.783\rho \sin(y_1)}, \tag{3.137}$$

where $y_1^2 = \rho^2 + (3.832)^2$ and for an infinitely long cylinder it is

$$2a \sin\theta_1 = \left[1 + \frac{\rho\pi}{2\hat{y}_1} Y_1(\hat{y}_1)\right], \tag{3.138}$$

with $\hat{y}_1^2 = \rho^2 + \pi^2$ and Y_1 being the modified cylindrical Bessel function.

3.6.4 Infinitely long cylinders: Normal incidence

The derivation of the scattering function in the EA for the case of an infinitely long cylinder closely follows the derivation for a sphere. Thus, the field inside an infinitely long cylinder is written as (Alvarez-Estrada et al. 1980, Sharma and Somerford 2006):

$$\Psi(\mathbf{b}) \approx \exp\left[ikx + \frac{ik}{2}\int_{-\sqrt{a^2-y^2}}^{-\infty} \left(m^2(x', y) - 1\right) dx'\right], \tag{3.139}$$

where $\mathbf{b}$ is a two-dimensional vector in a plane which is perpendicular to the axis of the cylinder (z-axis). The direction of propagation of the wave is taken to be along the x-axis. The characteristic size a is the radius of the cylinder.

Inserting (3.139) in (2.167) and making a small angle approximation, $\mathbf{q}.\mathbf{b} \approx qy$, the integration over x can be executed to yield:

$$T(\theta)_{EA} = -\frac{k}{2}\int_{-a}^{a} dy\; e^{iqy}\left[e^{i\chi_{EA}(y)} - 1\right], \tag{3.140}$$

where $q = -k\sin\theta$, and

$$\chi_{EA}(y) = \frac{k}{2}\int_{-\sqrt{a^2-y^2}}^{\sqrt{a^2-y^2}} \left[m^2(x', y) - 1\right] dx'. \tag{3.141}$$

For a homogeneous cylinder $m(\mathbf{r}) = m$. In addition, if a change of variable $y = a\sin\gamma$ is made, (3.140) can be recast as,

$$T(\theta)_{EA} = x\int_0^{\pi/2} d\gamma\; \cos\gamma \cos(\tilde{z}\sin\gamma)\left[1 - \exp(\tilde{\omega}_{EA}\cos\gamma)\right], \tag{3.142}$$

Here, $\tilde{z} = x \sin\theta$ and as before $\tilde{\omega}_{EA} = (i\tilde{\rho}_{EA} - \tilde{\rho}_{EA} \tan\beta_{EA})$. It can be easily verified that in the RGA and Fraunhofer diffraction limits, (3.142) yields the expected expressions. This confirms again, that the EA applies to the entire x region despite the original premise $x \gg 1$.

For $\theta = 0$, (3.142) yields analytic expressions. For an absorbing cylinder, it gives

$$T(0)_{EA} = \frac{x\pi}{2} L(\tilde{\omega}_{EA}), \tag{3.143}$$

and for a nonabsorbing cylinder, it gives

$$T(0)_{EA} = \frac{\pi x}{2} \left[i J_1(\rho_{EA}) + S_1(\rho_{EA}) \right]. \tag{3.144}$$

In (3.143), $L(\tilde{\omega}_{EA}) = I_1(\tilde{\omega}_{EA}) - L_1(\tilde{\omega}_{EA})$, where $I_1(\tilde{\omega}_{EA})$ is the modified Bessel function and $L_1(\tilde{\omega}_{EA})$ is the modified Struve function. $J_1(\rho_{EA})$ and $S_1(\rho_{EA})$ are, respectively, the Bessel and the Struve functions of order 1. The extinction efficiency can be obtained using the extinction theorem, $Q_{ext} = (2/x)\mathrm{Re}T(0)$. The derivation from the exact results can be found in Sharma (1989). Equation (3.144) has also been derived by starting from the exact series solutions (Sharma 1989).

For $\theta \neq 0$ and the real refractive index, a closed form analytic solution is obtained for the imaginary part of the $T(\theta)_{EA}$:

$$\mathrm{Im}\ T(\theta)_{EA} = -\frac{i\pi x \rho_{EA}}{2\hat{y}} J_1(\hat{y}). \tag{3.145}$$

The real part of $T(\theta)_{EA}$ can be expressed in a series form in terms of known functions (Sharma et al. 1997a):

$$\mathrm{Re}\ T(\theta)_{EA} = \frac{\pi x}{2} \sum_0^{\infty} \frac{(-1)^k}{2k!!} \left(\frac{z^2}{\rho_{EA}} \right)^k S_{k+1}(\rho_{EA}), \tag{3.146a}$$

for $\rho_{EA} > 1$ and

$$\mathrm{Re}\ T(\theta)_{EA} = \frac{x\rho_{EA}^2}{2} \sqrt{\frac{\pi}{2\tilde{z}}} \sum_0^{\infty} \frac{\Gamma(k+1)}{(2k+1)!} \left(\frac{2\rho_{EA}^2}{z} \right)^k J_{k+3/2}(z), \tag{3.146b}$$

for $\rho_{EA} < 1$. Again, the refractive index is real here too. Interesting analytic formulas for the positions of minima have been obtained by truncating these series appropriately (Sharma et al. 1997a).

Corrections, analogous to those for a sphere, have also been obtained (Sharma et al., 1981, Sharma 1993). A closed form analytic expression for the first-order correction to $T(0)_{EA}$ is ,

$$T(0)_{FCI} = T(0)_{EA} - \frac{\tilde{\rho}_{EA}^2}{4} \left[\left(\frac{4\tilde{\rho}_{EA}}{3} \, {}_1F_2(2, \frac{3}{2}, \frac{5}{2}; \frac{-\tilde{\rho}_{EA}^2}{4}) + \frac{i\pi}{2} S_0(\tilde{\rho}_{EA}) \right) \right.$$

$$+\frac{i\pi}{2}\left({}_1F_2(\frac{3}{2},\frac{1}{2},2;\frac{-\tilde{\rho}_{EA}^2}{4}) - J_0(\tilde{\rho}_{EA})\right)\Bigg]. \qquad (3.147)$$

In (3.147), $S_0(\tilde{\rho}_{EA})$ is the Struve function of order zero and ${}_1F_2$ is the generalized hypergeometric function.

For arbitrary θ, Di Marzio and Szajman (1992) have obtained the following series form expression for the first-order correction:

$$\text{Re } T(\theta)_{FCI} = \text{Re } T(\theta)_{EA} - \frac{\rho^2}{4}\sum_{i=0}^{\infty}\frac{(-1)^i\rho^{2i+1}}{(2i+1)!}\sum_{j=0}^{i}\binom{2i+1}{2j}\left(\frac{z}{\rho}\right)^{2j}$$

$$\times\left[\sum_{k=0}^{i-j}\frac{(-1)^k}{2j+2k+1}\binom{i-j}{k}\times\left(\frac{i-j+1}{i-j-k+1}-\frac{2j+2k+1}{2j+2k+3}\right)\right.$$

$$\left.+\frac{(-1)^{i-j+1}}{2i+3}\right], \qquad (3.148a)$$

$$\text{Im } T(\theta)_{FCI} = \text{Im } T(\theta)_{EA} - \frac{\pi\rho^2}{8}\sum_{i=0}^{\infty}\frac{(-1)^i}{(2i)!}\sum_{j=0}^{i}\binom{2i}{2j}\left[\rho^{2j}z^{2(i-j)} - z^{2j}\rho^{2(i-j)}\right]$$

$$\times(-1)^k\binom{j+1}{k}\frac{(2i-2j+2k-1)!!}{(2i-2j+2k)!!}, \qquad (3.148b)$$

where Im $T(\theta)$ is given by equation (3.145) and Re $T(\theta)$ is given by equation (3.146a).

3.6.5 Coated spheres

The problem of scattering of light by a coated particle has been addressed in the framework of the EA and the ADA by many researchers (see, for example, Morris and Jennings 1977, Aas 1984, Chen 1987, Lopatin and Sid'ko 1987, Zege and Kokhanovsky 1989, Sharma and Somerford 2006, among others). This geometry has been extensively employed to model many biological, atmospheric and environmental particles. If m_1, a_1 represent the relative refractive index with respect to the surrounding medium and radius, respectively, of the core and m_2, a_2 represent the corresponding quantities for the coating, the scattering function in the EA takes the form,

$$S(\theta)_{EA} = k^2\left[\int_0^{a_1} bdb\ J_0(qb)\left(1 - e^{ip_1\sqrt{a_1^2-b^2}+ip_2(\sqrt{a_2^2-b^2}-\sqrt{a_1^2-b^2})}\right)\right.$$

$$\left.+\int_{a_1}^{a_2} bdb\ J_0(qb)\left(1 - e^{ip_2\sqrt{a_2^2-b^2}}\right)\right], \qquad (3.149)$$

where $p_1 = k(m_1^2 - 1)$ and $p_2 = k(m_2^2 - 1)$. Applying the change of variable, $b = a_1 \sin\gamma_1 = a_2 \sin\gamma_2$, (3.109) can be recast as,

$$S(\theta)_{EA} = x_1^2 \int_0^{\pi/2} d\gamma_1 \, \frac{\sin 2\gamma_1}{2} J_0(qa_1 \sin\gamma_1)\left[1 - e^{i(p_1-p_2)a_1\cos\gamma_1 + ip_2a_2\cos\gamma_2}\right]$$

$$+x_2^2 \int_{\sin^{-1}(a_1/a_2)}^{\pi/2} d\gamma_2 \cos\gamma_2 \sin\gamma_2 J_0(qa_2 \sin\gamma_2)\left[1 - e^{ip_2a_2\cos\gamma_2}\right]. \quad (3.150)$$

Simple expressions have been obtained for special cases of thin and thick coatings (Morris and Jennings 1977, Aragon and Elwenspoek 1982, Bhandari 1986).

3.6.6 Spheroids

A formal exact solution for the scattering by a spheroid was suggested first by Moeglich (1927). But, a complete solution is credited to Asano and Yamamoto (1975). Like the Mie solution, this is also a complex infinite series solution. However, within the framework of the EA, the scattering function for a spheroid, $S_{oid}(a, b, m)$, can be shown to be related to the scattering function of a sphere $S_{ere}(a', m')$ as (Chen 1995, Chen and Yang 1996):

$$S_{oid}(a, b, m, \theta) = \frac{\alpha}{\beta^2} S_{ere}(a', m', \theta), \quad (3.151)$$

where a and b are semi-major and semi-minor axes of the spheroid and m is its refractive index. The radius and refractive index of the equivalent sphere are a' and m', respectively. These are related via the following relations,

$$a' = \frac{\beta}{e} a \quad \text{and} \quad m' = 1 + \frac{e}{\alpha\beta}(m-1), \quad (3.151a)$$

where $e = a/b$ is the aspect ratio of the spheroid,

$$\alpha = \sqrt{\cos^2\Theta + e^2 \sin^2\Theta}, \quad \beta = \sqrt{\alpha^2 \cos^2\Phi + \sin^2\Phi}, \quad (3.151b)$$

and Θ and Φ are related to θ_0, ϕ_0 and θ via the relations:

$$\cos\Phi = \frac{\cos(\theta/2)\cos\theta_0\cos\phi_0 - \sin(\theta/2)\cos\theta_0}{\sin\Theta}, \quad (3.151c)$$

$$\cos\Theta = \cos(\theta/2)\cos\theta_0 + \sin(\theta/2)\sin\theta_0\cos\phi_0. \quad (3.151d)$$

The coordinates (a, θ_0, φ_0) define the orientation of the spheroid in the laboratory frame.

The relationship (3.151) is found to be valid for exact scattering functions also if, for a prolate spheroid $kb^2/a \geq 4$ and for an oblate spheroid $ka^2/b \geq 4$

for $\theta \leq 30$ deg. Equation (3.151) has been tested for the GEA too. The agreement continues to be reasonable for scattering angles up to about 30 deg. For a prolate spheroid, the GEA works better when the direction of the incidence is normal to the major axis. For an oblate spheroid, it works better when the direction of incidence is parallel to the major axis.

3.6.7 Backscattering in the eikonal approximation

Although the EA was derived as a near forward scattering approximation, it successfully serves as a backward scattering approximation too. If it is assumed that the backscattering is a result of the single hard scattering, and not of the accumulation of a number of soft scatterings, the backscattering function can be written as (Saxon and Schiff 1957):

$$S(\pi)_{SS} = -i\frac{(1-m^2)x}{8}\left[e^{2ixm^2} + e^{-2ix}\right]. \tag{3.152}$$

The scattered intensity corresponding to (3.152) for a nonabsorbing sphere is,

$$i(\pi)_{SS} = \frac{x^2(m^2-1)^2}{16}\cos^2[x(m^2+1)]. \tag{3.153}$$

The positions of minima in the scattering pattern are then given by the relation $i(\pi)_{SS} = 0$, leading to the condition

$$x(m^2+1) = p\pi, \tag{3.154}$$

where p is an integer. The separation between two successive minima for a given m is thus related to refractive index through the relation

$$\Delta x = \frac{\pi}{m^2+1}. \tag{3.155}$$

The predicted separation has been found to be in good agreement with the predictions of the Mie theory.

Two and three hard scattering contributions can also be taken into consideration (Reading and Bassichis 1972, Sharma and Somerford 1994) to give:

$$S(\pi)^{I}_{SS} = \frac{x(m^2-1)^2}{8}\sqrt{\frac{x\pi}{\sqrt{2}}}e^{2ix\sqrt{2}+\frac{3ix}{\sqrt{2}}(m^2-1)-\frac{i\pi}{4}}, \tag{3.156}$$

and

$$S(\pi)^{II}_{SS} = -ix(m^2-1)^3\sqrt{6\pi x}e^{3ix+2ix(m^2-1)-\frac{i\pi}{4}}. \tag{3.157}$$

If the medium is highly absorbing the main contribution comes from the first reflection. In contrast, for a nonabsorbing dielectric sphere $|S(\pi)_{SS}|^2/|S(\pi)^{I}_{SS}|^2 = x(n_r^2-1)^2$, the contribution from two hard scattering events becomes important and may even dominate if $x > 1/|n_r^2-1|^2$.

A comparison of $i(\pi)_{SS}$ for nonabsorbing spheres of $n_r = 1.05$ and $5.0 \leq x \leq 50$ shows good agreement with exact results except for deeper minima. The positions of minima are reproduced quite accurately. The deeper minima predicted by $i(\pi)_{SS}$ get filled up by contributions from two hard scattering events to get closer to the exact values.

3.7 Anomalous diffraction approximation

Formally, the scattering function in the ADA is identical to that in the EA. The difference lies in that $(m^2 - 1)$ in the EA is replaced by $2(m - 1)$ in the ADA. The scattering function in the ADA can thus be written as:

$$S(\theta)_{ADA} = -k^2 \int d\mathbf{b} e^{i\mathbf{q}.\mathbf{b}} \left[e^{i\chi(\mathbf{b})_{ADA}} - 1 \right], \tag{3.158}$$

where

$$\chi(\mathbf{b})_{ADA} = k \int_{-\infty}^{\infty} \left[m(\mathbf{r}) - 1 \right] dz. \tag{3.159}$$

As $m \to 1$, the two approximations tend to the same limit. The conditions (3.108a) and (3.108b) define the validity domain of the ADA too. Formally, all results obtained for the EA are also valid for the ADA and vice versa.

An alternative way to write the ADA expressions is to encode phase shifts in terms of the chord length. If l is the geometrical path length of a given ray in passing through the particle, the phase shift suffered by it is $kl(m - 1)$. In terms of l and the projected area P, the extinction efficiency can be written as (Chylek and Klett 1991b):

$$Q_{ext}^{ADA} = \frac{2}{P} \int \int_P \left[1 - e^{ikl(m-1)} \right] dP, \tag{3.160}$$

using (3.158) and the extinction theorem. The absorption efficiency factor, using equation (3.124), becomes,

$$Q_{abs}^{ADA} = \frac{1}{P} \int \int_P \left[1 - e^{-kln_i} \right] dP. \tag{3.161}$$

In this representation, the computation of two quantities depend only on the morphology of the scatterer. Hence, it is much faster. For this reason, this representation of the ADA has been referred to as the ADr. That is, anomalous diffraction rapid.

3.7.1 Homogeneous spheres

For a homogeneous sphere, the integration over the azimuthal angle in (3.158) can be easily performed and the scattering function may be set as

$$S(\theta)_{ADA} = -k^2 \int_0^a b\ db\ J_0\Big(2kb\sin(\theta/2)\Big)\Big[\exp\Big(i\chi(b)_{ADA}\Big) - 1\Big], \quad (3.162)$$

where

$$\chi(b)_{ADA} = 2x(m-1)\sqrt{[1-(b^2/a^2)]} = \tilde{\rho}_{ADA}\sqrt{[1-(b^2/a^2)]}, \quad (3.163)$$

with

$$\tilde{\rho}_{ADA} = 2x(n_r - 1) + 2ixn_i \qquad \equiv \rho_{ADA} + i\kappa_{ADA}. \quad (3.164)$$

The quantity $\tilde{\rho}_{ADA}$ is also expressed as $\tilde{\rho}_{ADA}(1 + i\tan\beta_{ADA})$, where

$$\tan\beta_{ADA} = 2n_i x/\rho_{ADA}. \quad (3.165)$$

For forward scattering by a nonabsorbing sphere, the integration in (3.164) can be performed to yield a closed form analytic expression for $S(0)$. The extinction theorem then yields:

$$Q_{ext}^{ADA} = \left[2 - \frac{4}{\rho_{ADA}}\sin\rho_{ADA} + \frac{4}{\rho_{ADA}^2}(1-\cos\rho_{ADA})\right]. \quad (3.166)$$

When ρ is sufficiently large, the last term on the right-hand side of (3.166) may be neglected. The oscillations in the curve against λ are then essentially due to $\sin\rho/\rho$ and the diameter of the particle can be derived to be (Xu et al. 2004):

$$d = \frac{1}{n-1}\frac{\lambda_1\lambda_2}{\lambda_2 - \lambda_1}, \quad (3.167)$$

where λ_1 and λ_2 are the wavelengths corresponding to two adjacent extrema.

3.7.2 Edge effects

The extinction efficiency factors in the EA as well as in the ADA lead to the correct $x \to \infty$ limit of the extinction efficiency. However, the rate of approach to this limit is much faster than that predicted by the Mie theory. This difference is attributed to the effect at the edge of the particle. This contribution can be included in the extinction formula as an additional term (see, for example, Nussenzweig and Wiscombe 1980, Ackerman and Stephens 1987):

$$Q_{ext} = Q_{ext}^{ADA} + Q_{edge}, \quad (3.168)$$

where

$$Q_{edge} = 2x^{-2/3}. \quad (3.168a)$$

The edge corrections are essentially wave optics corrections to geometrical optics results. Therefore, these apply only to large particles.

An empirical recipe for the edge effect has been given by Klett (1984). The extinction efficiency factor, including edge effects, is expressed as,

$$Q_{ext}^{K} = \mathcal{D} Q_{ext}^{ADA}, \tag{3.169}$$

where $\mathcal{D} = 1.1 + (|m| - 1.2)/3$. This *ad hoc* prescription is valid for $|m| = 1 - 5$. The modification concurs closely with a correction term given in (3.127a).

3.7.3 Relationship with the Ramsauer approach

Ramsauer (1921) obtained a scattering formula in the context of nuclear scattering. When translated to optical scattering, it turns out to be essentially the same formula as the ADA (Louedec and Urban 2012):

$$S(\theta)_{RAA} = -k^2 \frac{1+\cos^2\theta}{2} \int_0^a b\, db\; J_0\Big(kb\sin\theta\Big)\Big[\exp\Big(i\chi(b)_{ADA}\Big) - 1\Big], \tag{3.170}$$

The Ramsauer phase function is then defined as,

$$\phi(\theta)_{RAA} = \frac{|S(\theta)_{RAA}|^2}{2\pi \int_0^{\pi} |S(\theta)_{RAA}|^2 \sin\theta d\theta}. \tag{3.170b}$$

If the scatterer is a cylinder of radius a and height $2a$ and if the radiation is incident along the axis of the cylinder, the phase shift suffered by the a ray in travelling through the cylinder is $\chi_{ADA} = 2ka(m-1)$. This is independent of b. Then, b integration in (3.170) can be performed to yield,

$$|S(\theta)_{RAA}|^2 = \frac{C_{sca}}{4\pi}\left[x\,(1+\cos\theta)\,\frac{2J_1(x\sin\theta)}{x\sin\theta}\right]^2. \tag{3.171}$$

Assuming that a form analogous to (3.171) holds for a sphere too, the phase function for a sphere can be cast as,

$$\phi(\theta)_{SRA} = \frac{1}{4\pi}\left[x\,(1+\cos\theta)\,\frac{2J_1(x\sin\theta)}{x\sin\theta}\right]^2. \tag{3.172}$$

The subscript SRA stands for simplified Ramsauer approximation. Note, that the scattering phase function (3.172) does not depend on the refractive index of the scatterer. Hence, it can be used as a first stage for developing models without knowledge of the refractive index.

Expressions for the total scattering cross section and the scattering efficiency factor are:

$$C_{sca}^{SRA} = 2\pi R^2\left[1 - 2\frac{\sin(4x\sin(\theta/2))}{4x\sin(\theta/2)} + \left(\frac{\sin(2x\sin(\theta/2))}{2x\sin(\theta/2)}\right)^2\right], \tag{3.173}$$

and

$$Q_{ext}^{SRA} = 2\left(1+\frac{m-1}{\rho_{ADA}}\right)^2\left[1-2\frac{\sin\,\rho_{ADA}}{\rho_{ADA}}+\left(\frac{\sin\,\rho_{ADA}}{\rho_{ADA}}\right)^2\right]. \tag{3.174}$$

While writing (3.174), additional assumptions have been made to introduce m dependence. The outside factor is a result of certain geometrical considerations.

Numerical comparisons for $m = 1.05,\ 1.33,\ 2.0$ show that the SRA phase function agrees reasonably. On the other hand, (3.173) and (3.174) show very good agreement with the Mie calculations.

3.7.4 X-ray scattering in the ADA

Draine and collaborators have considered the problem of X-ray scattering by interstellar dust particles (Draine and Allaf-Akbari 2006, Hoffman and Draine 2016). The conditions for the validity of the ADA in the X-ray domain can be seen to be

$$E > 60eV \quad \text{and } a > 0.035\ \mu m \times (60eV/E). \tag{3.175}$$

Hoffman and Draine (2016) have written a computer program in FORTRAN 95-"general geometry anomalous diffraction theory (GGADT)", which calculates the energy dependent scattering and absorption cross sections and the differential scattering cross section for grains of arbitrary geometry.

The results from GGADT have been used by Hoffman and Draine (2016) to compare extinction cross sections of five particle geometries to assess the effect of particle shapes. The five geometries are:

(i) A sphere,
(ii) An oblate spheroid,
(iii) A prolate spheroid,
(iv) A BA type aggregate and
(v) A BAM2 type aggregate.

The last two shapes are porous agglomerates. The porosity, $\mathcal{P}$, is defined by the relation,

$$a_{EE}^3(1-\mathcal{P}) = a_{eff}^3, \tag{3.176}$$

where $(4\pi/3)a_{EE}^3$ is the volume of the equivalent sphere. The effective particle size, a_{eff}, is described through the relation,

$$a_{eff} = \left(\frac{3\mathcal{V}_s}{4\pi}\right)^{1/3}, \tag{3.176a}$$

where $\mathcal{V}_s$ is the volume of the solid material. Thus, $a_{EE} = a_{eff}(1-\mathcal{P})^{1/3}$. The dielectric function of the grains is taken to be that for $MgFeSiO_4$. A detailed description of aggregates can be found in Shen et al. (2008).

For spherical grains, computations from GGADT were indistinguishable from the Mie theory. For spheroids, the orientation averaged extinction cross sections did not differ much from spherical dust grains. However, the absorption and scattering cross sections of random aggregates differed appreciably from that of a sphere.

3.7.5 Long cylinders: Oblique incidence

Let Ψ be the angle that the incident wave makes with the axis of the cylinder. Perpendicular incidence corresponds to $\Psi = \pi/2$. The phase shift suffered by the ray in passing through the cylinder is simply $1/\sin\Psi$ times that for the perpendicular incidence (Stephens 1984). The scattering function and the extinction efficiency factor thus become,

$$T(\theta)_{ADA} = x\int_0^{\pi/2} d\gamma\ \cos\gamma\ \cos(z\sin\gamma)\Big[1-\exp(i\tilde{\rho}_{ADA}\cos\gamma/\sin\Psi)\Big], \tag{3.177}$$

and

$$Q_{ext}^{ADA} = \pi\ \mathrm{Re}\left[\frac{L(\tilde{\omega}_{ADA}}{\sin\Psi}\right], \tag{3.177a}$$

respectively. In (3.177a) $\tilde{\omega}_{ADA} = i\tilde{\rho}_{ADA} - \tilde{\rho}_{ADA}\tan\beta_{ADA}$. The same result has been re-derived by Sharma (1989) starting from the exact analytic solution of the problem.

Extinction efficiency factors corresponding to (3.177a) and (3.143) have been contrasted with the exact results by Stephens (1984) and by Fournier and Evans (1996), respectively, for real refractive index values. While the agreement is found to be good for normal incidence, the maxima and the minima in Q_{ext}^{ADA} versus the x curve are out of phase with the exact curves. The deviation increases with increasing obliqueness.

An extended ADA (EADA) wherein

$$\tilde{\rho}_{ADA} = 2x[(m^2-\cos^2\Psi)^{1/2} - \sin\Psi], \tag{3.178}$$

has been suggested and investigated by Fournier and Evans (1996). This is the analogue of the GEA for a sphere. The contribution of the edge term is also considered:

$$Q_{edge} = \frac{0.996130}{(x^{2/3}+x_{crit})\sin^{2/3}\Psi}, \tag{3.179}$$

where

$$x_{crit} = \frac{3.6}{r|(m^2-\cos^2\Psi)^{1/2}-\sin\Psi|}. \tag{3.179a}$$

The EADA improves over the ADA results dramatically.

Closed form expressions for extinction and absorption efficiency factors have been obtained in the ADA by Cross and Latimer (1970) too at normal as well as at oblique incidence.

3.7.6 Long elliptic cylinders

The ADA or the EA expressions for infinitely long circular cylinders can be easily extended to the infinitely long cylinders with elliptic cross sections. Formally, the extinction efficiency factor remains unaltered:

$$Q_{ext}^{ADA} = \pi[S_1(\tilde{\rho}_{ADA}) + iJ_1(\tilde{\rho}_{ADA})]. \qquad (3.180)$$

The phase shift term, $\tilde{\rho}_{ADA}$, however, changes to

$$\tilde{\rho}_{ADA} = \frac{2(m-1)rb}{p \sin \Psi}, \qquad (3.181a)$$

where, because of the elliptic cross section,

$$p = (\cos^2 v + r^2 \sin^2 v), \qquad (3.181b)$$

with v as the angle between the semi-major axis of the ellipse and the plane defined by the direction of the incident radiation and the long axis of the cylinder. The quantity r is the ratio of the semi-major axis a to the semi-minor axis b. Note that Q_{ext}^{ADA} for an elliptic cylinder is formally the same as a circular cylinder but with x replaced by x/p.

The edge term is derived to be

$$Q_{edge} = \frac{0.993160}{p^2} \left[\frac{r}{b \sin \Psi} \right]^{2/3}. \qquad (3.182)$$

Numerical estimates of the validity of (3.102) do not seem to have been performed.

3.7.7 Spheroids

The possibility of using an equivalent sphere description for the case of a spheroid has been examined by Chen et al. (2003, 2004). Three modes were investigated. (i) The magnetic vector is perpendicular to the major axis of the spheroid and the electric vector is parallel to it (TM mode). (ii) The electric vector is perpendicular and the magnetic vector is parallel to the major axis (TE mode). (iii) The electric vector as well as the magnetic vector is perpendicular to the major axis of the spheroid (TEM mode). The extinction efficiency factor in the TE and TM mode is:

$$Q_{ext}^{ADA} = \frac{4}{\pi ab}\left[\frac{\pi ab}{2}\left(1 - 2m\frac{\sin\tilde{\rho}_b}{\tilde{\rho}_b} + 4m\frac{\sin^2(\tilde{\rho}_b/2)}{\tilde{\rho}_b} + \frac{\pi ab}{2}\frac{k(ab^2)^{1/3}}{2}\right)^{-2/3}\right], \tag{3.183}$$

where $\tilde{\rho}_b = ka(m-1)$. The validity range is given by

$$\frac{16b}{\pi^2\lambda}\frac{m^2-1}{m}\left(1-\frac{b}{a}\right)\left(1+\frac{a^2}{b^2}\right)^{-1} < 1. \tag{3.184}$$

For TEM mode,

$$Q_{ext}^{ADA} = \frac{4}{\pi b^2}\left[\frac{\pi b^2}{2}\left(1 - 2m\frac{\sin\tilde{\rho}_a}{\tilde{\rho}_a} + 4m\frac{\sin^2(\tilde{\rho}_a/2)}{\tilde{\rho}_a} + \frac{\pi ab}{2}\frac{k(ab^2)^{1/3}}{2}\right)^{-2/3}\right], \tag{3.185}$$

where $\tilde{\rho}_a = ka(m-1)$. The validity range is,

$$\frac{16a}{\pi^2\lambda}\frac{m^2-1}{m}\left(1-\frac{a}{b}\right)\left(1+\frac{b^2}{a^2}\right)^{-1} < 1. \tag{3.186}$$

These restrictions are in addition to the usual conditions required for the validity of the ADA.

3.7.8 Ellipsoids

Scattering by an ellipsoid in the ADA has been examined, among others, by Lind and Greenberg (1966), Latimer (1980), Paramonov et.al. (1986), Lopatin and Sid'ko (1988), Streekstra et al. (1993), Paramonov (1994), Streekstra (1994). An orientation of the ellipsoid which is of particular interest in ektacytometric applications is when $\mathbf{k}_i$ is along the third axis of the ellipsoid. The scattering function may then be written as (see, for example, Streekstra et al. 1993, Streekstra 1994):

$$S(\theta) = k^2A^2\int_0^{\pi/2}\left[1 - e^{i\tilde{\rho}\sin\gamma}\right]J_0(kA\cos\gamma\sin\theta)\cos\gamma\sin\gamma\, d\gamma \tag{3.187}$$

where $A = \sqrt{ab}$ with a as the long axis and b as the short axis of the cross-sectional ellipse and

$$\tilde{\rho} = 2kc(m-1), \tag{3.188}$$

with c as the length of the third axis of the ellipsoid parallel to the direction of the incident wave vector. Numerically, (3.187) has been found to be highly accurate within the scattering angle of about 15 degrees in the context of scattering of light by the ellipsoidal-shaped red blood cells.

An equivalent sphere approach gives the following diameter for the equivalent sphere (Kokhanovsky 2004):

$$h = \frac{2abc}{\sqrt{(bc\ \cos\mu_1)^2 + (ac\ \cos\mu_2)^2 + (ab\ \cos\mu_3)^2}} = \frac{3\mathcal{V}}{P}, \tag{3.189}$$

where a, b, c are semi-axes of the ellipsoidal particle, and μ_1, μ_2, μ_3 are angles which the incident radiation makes with the coordinate axes x, y, z respectively. The symbol $\mathcal{V}$ represents the volume of the particle, P is its geometrical cross section and h is the maximal length of a ray inside the particle.

3.7.9 Layered particles

The formal expression for the scattering function for a coated sphere also remains the same as in the EA (eq. 3.149). The change is that in the ADA $p_1 = 2k(m_1 - 1)$ and $p_2 = 2k(m_2 - 1)$. The extinction efficiency factor can then be expressed as (Quirantes and Bernard 2004):

$$Q_{ext}^{ADA} = 2 - 4\tilde{z}e^{-\tilde{z}\rho_1 \tan\beta_1}\frac{\cos\beta_1}{\rho_1}\sin(\tilde{z}\rho_1 - \beta_1)$$

$$4\frac{\cos\beta_1}{\rho_1}^2 e^{-\tilde{z}\rho_1 \tan\beta_1}\cos(\tilde{z}\rho_1 - 2\beta_1) + 4\left(\frac{\cos\beta_1}{\rho_1}\right)^2\cos(2\beta_1)$$

$$4\frac{\cos\beta_1}{\rho_1}e^{-\rho_1 \tan\beta_2}\sin(\rho_2 - 2\beta_2) + 4\frac{\cos\beta_2}{\rho_2}e^{-\tilde{z}\rho_2 \tan\beta_2}\sin(\tilde{z}\rho_2 - 2\beta_2)$$

$$4\frac{\cos\beta_2}{\rho_2}^2 e^{-\rho_2 \tan\beta_2}\sin(\rho_2 - 2\beta_2) + 4\frac{\cos\beta_2}{\rho_2}^2 \tilde{z}e^{-\tilde{z}\rho_2 \tan\beta_2}\sin(\tilde{z}\rho_2 - 2\beta_2), \tag{3.190}$$

and

$$Q_{abs}^{ADA} = 1 + \frac{\tilde{z}e^{-2\kappa_1\tilde{z}}}{\kappa_1} + \frac{e^{-2\kappa_1\tilde{z}} - 1}{\kappa_1^2} + \frac{e^{-2\kappa_2} - \tilde{z}e^{-2\kappa_2\tilde{z}}}{\kappa_2} + \frac{e^{-2\kappa_2} - \tilde{z}e^{-2\kappa_2\tilde{z}}}{2\kappa_2}, \tag{3.191}$$

where

$$\tilde{z} = (1 - \xi^2), \quad \rho_1 = 2x_2(n_2 - 1), \quad \rho_2 = 2x_2\left[\xi n_1 + (1 - \xi)n_2 - 1\right], \tag{3.192a}$$

$$\kappa_1 = 2x_2 n_{2i}, \quad \xi = \frac{a_1}{a_2}, \quad \kappa_2 = 2x_2\left[\xi n_{1i} + (1 - \xi)n_{2i}\right], \tag{3.192b}$$

$$\tan\beta_1 = \frac{n_{2i}}{n_2 - 1}, \quad \tan\beta_2 = \frac{\xi n_{1i} + (1 - \xi)n_{2i}}{\xi n_1 + (1 - \xi)n_2 + 1}. \tag{3.192c}$$

For $\xi = 1$, the above results reduce to that for a homogeneous sphere. Closed form analytic expressions in terms of known functions for a coated sphere and a hollow sphere in the ADA can be found in Aas (1984) too.

The application of the ADA to multilayered particles is a straightforward extension of the procedure followed for a coated particle. The validity conditions now need to be satisfied in each layer. Theoretical expressions for efficiencies and scattering function are given in the book by Kokhanovsky (2004) for layered spheres and ellipsoids.

3.7.10 Other shapes

Chylek and Klett (1991a, 1991b) obtained general equations for Q_{ext} and Q_{abs} in the framework of the ADA for light scattering by a column with either a triangular, trapezoidal, hexagonal, or polygonal base. Analytic expressions for arbitrary polygonal bases have also been obtained by Heffels et al.(1995). Sun and Fu (1999) have derived analytic expressions for Q_{ext} and Q_{abs} for arbitrarily oriented hexagonal particles. Analytic results can also be obtained for plates and needles (Aas 1984, Chylek and Klett 1991b, Chylek and Videen 1994).

For a cube, Flatau (1992) and Maslowska et al. (1994) have examined ADA numerically against the DDA. Extinction and absorption efficiency factors were studied for: (i) edge incidence and (ii) side incidence. Side incidence results exhibit fairly good agreement between the two methods. The error is on the same order as that observed between Mie and the ADA for a sphere. For the edge-on incidence, the deviations from exact results are larger.

Rysakov (2004, 2006) has extended the basic concept of the RGA to the realm of the WKB approximation to deal with scattering and absorption of nonabsorbing soft particles of arbitrary shape and size. Relevant expressions for the sphere, cylinder and parallelepiped are given. Numerical calculations show that the errors are on the same order as that in the ADA.

3.8 WKB approximation

The WKB approximation (WKBA) was founded by Carlini (1817) and developed by Liouville (1837) and Green (1837) (see, for example, Malinka 2015). It was applied in optics by Rayleigh (1912) and Gans (1915) and in quantum mechanics by Brillouin (1926), Kramers (1926) and Wentzel (1926). In the context of electromagnetic wave scattering, it has been described in details by Saxon (1955). The condition for its validity is,

$$|m-1| \ll 1, \tag{3.193}$$

and can be obtained from the ADA by multiplying it by a factor $2/(m+1)$ (Klett and Suderland 1992, Malinka 2015).

A two wave WKBA has been studied in detail by Klett and Sutherland (1992), which assumes the field inside the particle as,

$$\Psi(\mathbf{r}) = e^{ik(m-1)\sqrt{a^2-b^2}}\left(e^{ikmz} + \frac{m-1}{m+1}e^{ikm(2\sqrt{a^2-b^2}-z)}\right). \tag{3.194}$$

The first term on the right-hand side of (3.194) is the standard ADA. The second term allows for the effect of reflection. When substituted in (3.107), it leads to following expression for the phase function for the scattering of unpolarized light by a sphere:

$$\phi(\theta) = \frac{2(1+\cos^2\theta)|H_1 + \exp(i\rho_1)RH_2|^2}{\int_0^\pi (1+\cos^2\theta)|H_1 + \exp(i\rho_1)RH_2|^2 \sin\theta d\theta}, \tag{3.195}$$

where

$$H_1(\theta) = \int_0^1 ydyJ_0(yx\sin\theta)\sin[x(m-\cos\theta)\sqrt{1-y^2}]\exp[ix(m-1)\sqrt{1-y^2}], \tag{3.195a}$$

$$H_2(\theta) = H_1(\pi - \theta), \quad R = \frac{1-m}{m+1}, \tag{3.195b}$$

and $\rho_1 = 2mx$.

The accuracy of the two wave WKBA has been examined by Jones et al. (1996). They showed that the two wave WKBA was superior to the conventional WKBA and the RGDA. The angular range of its applicability was found to be very limited beyond $\theta > 90$ deg. At these angles, the errors mostly exceeded 10%.

The backscatter efficiency in this approximation has been expressed as:

$$Q_{back} = \frac{4}{\pi}x^2\left|\frac{m-1}{m+1}\right|^2 |I_1 + \exp(i\rho_1)I_2|^2, \tag{3.196}$$

where

$$I_1 = \frac{1}{2}\left[\frac{i}{\rho_1^2} - (i+\rho_1)\frac{e^{i\rho_1}}{\rho_1^2} - \frac{i}{\rho_2^2} + (i-\rho_2)\frac{e^{-i\rho_2}}{\rho_2^2}\right], \tag{3.196a}$$

and

$$I_2 = \frac{i}{2}\left[(i\rho_3 - 1)\frac{e^{i\rho_3}}{\rho_3^2} + \frac{1}{\rho_3^2} + \frac{1}{2}\right], \tag{3.196b}$$

with $\rho_1 = 2xm$, $\rho_2 = 2x$ and $\rho_3 = 2x(m-1)$. The approximation has been found to be good for moderately soft particles. When substantial absorption is present the results can be improved considerably by using a multiplicative factor,

$$\left[1 + [1 - \exp(-n'x)](2\sqrt{\pi} - 1)\right]^2. \tag{3.197}$$

The fine structure of the backscatter however, cannot be improved.

The work of Klett and Sutherland has also been extended to the differential scattering cross sections for any particle shape with a given orientation and illuminated by an unpolarized light (Gruy 2014).

3.9 Perelman approximation

Perelman (1978, 1991) has given two closed form expressions for soft scatterers: (1) the simplest form of the Perelman approximation (PA) and (ii) the main form of the Perelman approximation (MPA). It is also known as the S-(soft particle) approximation.

3.9.1 Homogeneous spheres

For a homogeneous sphere this approximation is achieved by first casting the Mie coefficients a_n and b_n, given in (2,28) and (2.29), in the following form

$$a_n = \frac{h_{1n}}{h_{1n} + ih_{3n}}, \qquad b_n = \frac{h_{2n}}{h_{2n} + ih_{4n}}. \tag{3.198}$$

In the limit $m \to 1$ the denominators in (3.198) can be approximated as

$$h_{1n} + ih_{3n} \sim h_{2n} + ih_{4n} = i|m|^{-1/2}. \tag{3.199}$$

The summation over n in the expressions for $S_1(\theta)$ and $S_2(\theta)$ can then be performed for small angle scattering (Perelman 1991, Sharma and Somerford 2006):

$$S_1(\theta)_{PA} \simeq -2i(m+1)m|m|^{-1/2}x^2\left[\frac{\psi_1(\rho/2)}{\rho} - 2(1-\mu)mx^2\frac{\psi_2(\rho/2)}{\rho^2}\right], \tag{3.200a}$$

and

$$S_2(\theta)_{PA} \simeq -2i(m+1)m|m|^{-1/2}x^2\left[\frac{\mu\psi_1(\rho/2)}{\rho} - 2(1-\mu)mx^2\frac{\psi_2(\rho/2)}{\rho^2}\right]. \tag{3.200b}$$

Terms up to order θ^4 are retained in the expansions of π_n and τ_n.

The MPA is obtained by recasting (3.198) in the form,

$$a_n = \frac{h_{1n}(h_{1n} - ih_{3n})}{h_{1n}^2 + h_{3n}^2}; \qquad b_n = \frac{h_{2n}(h_{2n} - ih_{4n})}{h_{2n}^2 + h_{4n}^2}. \tag{3.201}$$

and approximating,

$$h_{1n}^2 + h_{3n}^2 \sim h_{2n}^2 + h_{4n}^2 = |m|. \tag{3.202}$$

The resulting series for $S(0)$ can be summed to yield (Perelman 1991):

$$S(0)^{MPA} = \frac{x^2}{8|m|}\left[(m^2+1)^2 + \frac{\omega(m,\rho) - \omega(-m,-R)}{2m}\right], \tag{3.203}$$

where

$$\omega(m,z) = [a(m) + a_0(m)z^2]ei(z) - ia_1(m)e_1(z) + a_2(m)e_2(z),$$

$$a(m) = (m^2-1)^2(m^2+1), \qquad a_0(m) = -2(m^2-1)^2(m-1)^2,$$

$$a_1 = (m+1)^2(m^4 - 2m^3 - 2m^2 - 2m + 1), \qquad a_2(m) = -a_0(m) - a_1(m),$$

$$ei(z) = \int_0^z dt\big(1 - \exp(-it)\big)/t, \qquad e_1(z) = \exp(-iz)/z,$$

$$e_2(z) = \big(1 - \exp(-iz)\big)/z^2, \tag{3.203a}$$

and $R = 2x(m+1)$. The above result holds for all values of $x > 0$ and $m = n_r + in_i$. For nonforward angles, the resulting series can still be summed, but only for small scattering angles.

The extinction efficiency for a nonabsorbing sphere can then be obtained as:

$$Q_{ext}^{MPA} = b_1\left[Q_h(\rho) - \left(\frac{m-1}{m+1}\right)^2 Q_h(R)\right] + b_2\left[Q_1(\rho) - Q_1(R) - \int_\rho^R \frac{1-\cos t}{t}dt\right]$$
$$+ \frac{(m-1)^2}{4m^2x^2}\left[Q_2(R) - Q_2(\rho) + \frac{1}{2}\int_\rho^R \frac{1-\cos t}{t}dt\right] + \frac{1}{2mx^2}Q_3(m,x), \tag{3.204}$$

where

$$b_1 = \frac{(m+1)^2(m^4 + 6m^2 + 1)}{32m^2}, \qquad b_2 = \frac{(m^2+1)(m^2-1)^2}{4m^2},$$

$$Q_1(\omega) = \frac{2(1-\cos\omega)}{\omega^2} + \frac{2\sin\omega}{\omega},$$

$$Q_2(\omega) = \frac{3\cos\omega}{8} + \frac{\sin\omega}{8} + \frac{\omega^2}{16},$$

$$Q_3(\omega) = (m-1)^2(\cos R - 1) + (m+1)^2(\cos\rho - 1),$$

and

$$Q_h(\rho) = Q_{ext}^{ADA}(\rho_{ADA}) = 2 - 4\frac{\sin\rho}{\rho} + 4\frac{1-\cos\rho}{\rho^2}. \qquad (3.204a)$$

An integral representation for the extinction efficiency factor (Granovskii and Ston 1994a, 1994b) is:

$$Q_{ext}^{MPA} = mx^4(m^2-1)^2 \int_{-1}^{1} dt(1+t^2)g^2(x\omega), \qquad (3.205)$$

where

$$g = \frac{\omega\cos\omega - \sin\omega}{\omega^3}; \quad \text{and} \quad \omega(t) = (1+m^2-2mt)^{1/2}. \qquad (3.205a)$$

The results from Q_{ext}^{MPA} agree with the exact results to within 5% in the m range $1.1 \leq m \leq 1.5$ and the x range $0 \leq x \leq x(m)$, if

$$x(m) \approx 290m^{-11},$$

where $x(m)$ gives the upper limit of x for a given m.

Perelman and Voshchinnikov (2002) have suggested an empirical improvement in the MPA:

$$Q_{ext}^{PVA} = \left[1 - \frac{S(m)-2}{S(m)}\exp\left(-\frac{0.01\exp(4m)}{u}\right)\right] Q_{ext}^{MPA}, \qquad (3.206)$$

where

$$S(m) = \frac{(m^2-1)^2}{2m} + \frac{(m^2+1)^2(m^2-1)^2}{4m^2}\ln\left|\frac{m-1}{m+1}\right|, \qquad (3.206a)$$

and $u = x/x(m)$.

For absorption efficiency, MPA gives the following expression (Perelman 1994) for weakly absorbing spheres,

$$Q_{abs}^{MPA} = 4n_r|m|^2 S(\kappa), \qquad (3.207)$$

where

$$S(\kappa) = \frac{\kappa\cosh\kappa - \sinh\kappa}{\kappa^2}, \qquad (3.208)$$

and $\kappa = 2n_i x$.

Comparison of extinction efficiency factors in the PVA and the ADA with the Mie results show that for very small x, the errors are greater than those in the RGA. For larger phase shifts, the PVA predicts very well the positions as well as the amplitudes of undulations in the extinction curve. It constitutes an improvement over the ADA. The error dependence of $x(m)$ has been estimated to be: (i) $x(m) \leq 240m^{-15.6}$ if the error will be less than 1%, (ii) $x(m) \leq 160m^{-9.30}$ if the error will be less than 5%, and (iii) $x(m) \leq 270m^{-9.99}$ if the error will be less than 10%.

3.9.2 The scalar Perelman approximation

If the approximation for b_n in (3.202) is implemented in (2.44), a straightforward algebraic manipulation leads to following expression for the forward scattering function:

$$S(0)_{SPA} = \frac{R^2}{64}\left[2 + \frac{4[1-\exp(i\rho)]}{\rho^2} + \frac{4i}{\rho}\exp(i\rho)\right]$$

$$-\frac{\rho^2}{64}\left[2 + \frac{4[1-\exp(-iR)]}{R^2} - \frac{4i}{R}\exp(-iR)\right]. \qquad (3.209)$$

The subscript SPA refers to scalar Perelman approximation. Clearly, it constitutes considerable simplification over the MPA. The first term on the right-hand side of (3.209) is nothing but $S(0)_{ADA}$ for a sphere, multiplied by a simple m and x dependent factor. The second term on the right-hand side of (3.209) can be obtained from the first term by replacing ρ by $-R$ and R by ρ. For a nonabsorbing sphere, the extinction efficiency factor then becomes,

$$Q^{ext}_{SPA} = \frac{(m+1)^2}{4}\left[Q_h(\rho) - \frac{\rho^2}{R^2}Q_h(-R)\right]. \qquad (3.210)$$

As $x \to \infty$ for a fixed ρ, $m \to 1$. In this limit, the right-hand side of (3.210) is nothing more than the $S(0)_{ADA}$.

Numerical comparison of the MPA and the SPA has been performed by Roy and Sharma (1996). Subsequent to this comparison, they proposed a modified SPA (MSPA) which gives,

$$Q^{ext}_{MSPA} = Q^{SPA}_{ext} - \phi(m-1) \qquad (3.211)$$

where

$$\phi(m-1) = \frac{1}{25}(m-1) + 5(m-1)^2 - 12(m-1)^3 - 2(m-1)^4. \qquad (3.211a)$$

The MSPA retains the simplicity of scalar approximation while being as accurate as the MPA.

3.9.3 Infinitely long cylinders

A procedure, analogous to that applied to the sphere, can be used for infinitely long cylinders too. For perpendicular incidence, it leads to following expressions for the scattering functions (Sharma et al. 1997b):

$$T(\theta)^{TMWS}_{PA} = \frac{-i\pi 2x(m^2-1)J_1(x\omega(\cos\theta))}{4\omega(\cos\theta)}, \qquad (3.212a)$$

and

$$T(\theta)^{TEWS}_{PA} = \cos\theta\, T(\theta)^{TMWS}_{PA}, \qquad (3.212b)$$

where $\omega(\cos\theta)$ is defined via the relation (3.205a). The MPA, for the scattering by an infinite cylinder yields,

$$T(\theta)^{TMWS}_{MPA} = \frac{\pi x^2(m^2-1)^2}{8}\int_0^{2\pi} d\phi \frac{J_1(x\omega(\cos(\theta-\phi))H_1(x\omega(\cos\phi))}{\omega(\cos\theta-\phi)\omega(\cos(\phi))}, \tag{3.213a}$$

$$T(\theta)^{TEWS}_{MPA} = \cos\theta \; T(\theta)^{TMWS}_{MPA}, \tag{3.213b}$$

where $H_1 = J_1 - iN_1$.

Employing the extinction theorem, the extinction efficiency factor in the MPA takes the form:

$$Q^{MPA}_{ext} = \frac{\pi(m^2-1)^2 x}{4}\int_0^{2\pi} d\phi \frac{J_1^2(x\omega(\cos\phi))}{\omega^2(\cos\phi)}. \tag{3.214}$$

The errors in Q^{MPA}_{ext} for a sphere are less than 5% for all values of x if $m \leq 1.6$, except possibly between $0 \leq x \leq 2$. The maximum error in Q^{MPA}_{ext} for an infinitely long homogeneous cylinder at perpendicular incidence for $m \leq 1.05$ and $x > 2.0$ is less than 2.27% (Sharma et al. 1997b). Typical comparison of Q_{ext} versus size parameter x curves obtained using the MPA, the SPA and the MSPA show that the accuracy of the MPA and the MSPA are nearly identical and predictions of the MPA and the MSPA are more precise than the SPA. For x close to 1, no approximation performs well except MPA. This is as expected because only the Q^{MPA}_{ext} reduces to Q^{ext}_{RGDA} for $x \leq 1$.

For scattered intensity, the positions of the maxima and minima as well as their amplitudes are reproduced quite decently for the first few extrema in the MPA. The MPA and the MSPA are more precise than the SPA. As expected, the errors increase as m increases.

3.10 Hart and Montroll approximation

3.10.1 Homogeneous spheres

The Hart and Montroll approximation (HMA) (Hart and Montroll 1951, Montroll and Hart 1951) considers denominators of the Mie coefficients in the limit $x \gg n$, n being the summation index in the Mie series. The resulting Mie series can then be summed.

When $x \gg n$, it can be shown that by employing the asymptotic expansions for Bessel and Hankel functions, the denominators of the Mie coefficients reduce to:

$$h_{1n} + ih_{3n} \sim \frac{i(m+1)m}{2}\exp(-i\rho/2)\big[1-(-1)^n\tilde{r}\;\exp(2imx)\big], \tag{3.215a}$$

and

$$h_{2n} + ih_{4n} \sim \frac{+i(m+1)m}{2} \exp(-i\rho/2)\big[1 + (-1)^n \tilde{r}\ \exp(2imx)\big], \quad (3.215b)$$

where $\tilde{r} = (m-1)/(m+1)$. It is interesting to note that in the limit $m \to 1$,

$$h_{1n} + ih_{3n} \sim h_{2n} + ih_{4n} = im. \quad (3.215c)$$

Note, this this approximation is then essentially identical to the PA. This implies that the HMA is valid under both sets of conditions, namely,

$$|m-1| \ll 1, \quad \text{and} \quad x \gg n. \quad (3.216)$$

By inserting (3.215) in the Mie series, the $S_1(\theta)$ and $S_2(\theta)$ can be summed to yield,

$$S_1(\theta)_{HMA} = \frac{-i\pi(m-1)\exp(i\rho/2)}{m^{1/2}\sin\theta[1-\tilde{r}^2\exp(4imx)]}\big[G_1(\theta) + \tilde{r}\ \exp(2imx)G_1(\pi-\theta)\big], \quad (3.217a)$$

and

$$S_2(\theta)_{HMA} = \frac{i\pi(m-1)\exp(i\rho/2)}{m^{1/2}\sin\theta[1-\tilde{r}^2\exp(4imx)]}\big[G_2(\theta) + \tilde{r}\ \exp(2imx)G_2(\pi-\theta)\big], \quad (3.217b)$$

where G_1 and G_2 are given by equations,

$$\begin{pmatrix} G_1 \\ G_2 \end{pmatrix} = \left(\frac{2m}{\pi}\right)^{1/2} \frac{mJ_{3/2}(x\omega(\cos\theta))\sin\theta}{(x\omega(\cos\theta))^{3/2}} \begin{pmatrix} 1 \\ \cos\theta \end{pmatrix}, \quad (3.218)$$

with $\omega(\cos\theta)$ as defined in (3.205a). In the corresponding expression for $G_J(\pi-\theta)$, $\omega(\cos\theta)$ is replaced by $\omega(-\cos\theta)$. Neglecting the terms of relative order r or higher, the extinction efficiency factor in the HMA can be cast as:

$$Q_{ext}^{HMA} = \frac{\pi x^2(m-1)^2}{2m}\Big[(m^2+6m^2+1)\Delta_1 - \frac{2(m^2-1)}{x^2}\Delta_2 + x^{-4}\Delta_3\Big], \quad (3.219)$$

where, the explicit expressions for Δ_1, Δ_2 and Δ_3, are (Sharma and Somerford 1996):

$$\Delta_1 = \frac{1}{\pi\rho^2}\Big[Q_h(\rho) - \frac{\rho^2}{r^2}Q_h(R)\Big], \quad (3.219a)$$

$$\Delta_2 = \frac{1}{\pi}\Big[Q_1(R) + c(R) - Q_1(\rho) - c(\rho)\Big] + \frac{4}{\pi}\Big[\frac{\cos R - 1}{R^2} - \frac{\cos\rho - 1}{\rho^2}\Big], \quad (3.219b)$$

and

$$\Delta_3 = \frac{2}{\pi}\Big[Q_1(R) - Q_1(\rho) + \frac{1}{2}c(R) - \frac{1}{2}c(\rho) + \cos R - \cos\rho\Big]. \quad (3.219c)$$

It may then be easily verified that (Sharma and Somerford 1996),

$$Q_{ext}^{HMA} = \frac{4m}{(m+1)^2} Q_{ext}^{MPA}. \qquad (3.220)$$

The factor $4m/(m+1)^2$ is very close to 1 for $|m-1| \ll 1$. For example, for $m = 1.05$ and $m = 1.10$ its value is 0.9994 and 0.9975, respectively. If one takes the limit $m \to 1$ for a fixed ρ of (3.219), the limiting formula is indeed the extinction efficiency in the ADA.

The fact that HMA holds for $x \gg n$ implies that its main contribution comes from the rays near the central ray. In fact, van de Hulst (1957) has suggested the name "central-incidence approximation" for it.

For scalar waves, van de Hulst (1957) casts the scattering efficiency in the following form,

$$Q_{sca} = \pi x^2 m(m-1)^2 \Big[\Phi(mx - x) - \Phi(mx + m)\Big], \qquad (3.221)$$

where

$$\Phi(u) = \frac{2}{\pi u^2}\left[1 - \frac{2\sin 2u}{2u} + \frac{2(1-\cos 2u)}{2u}\right]. \qquad (3.222)$$

In arriving at (3.221), terms of order $\tilde{r}$ have been neglected.

3.10.2 Infinitely long cylinders: Normal incidence

When $x \gg n$, it can be easily shown that by employing an asymptotic approximation for the Hankel and Bessel function, one can write

$$h_{1n} + ih_{3n} \sim \left(\frac{1}{xy}\right)^{1/2} \frac{i(m+1)}{\pi} \exp(\frac{-i\rho}{2}) \Big[1 - i\tilde{r}(-1)^n \exp(+2imx)\Big], \qquad (3.223)$$

$$h_{2n} + ih_{4n} \sim \left(\frac{1}{xy}\right)^{1/2} \frac{i(m+1)}{\pi} \exp(\frac{-i\rho}{2}) \Big[1 + i\tilde{r}(-1)^n \exp(+2imx)\Big], \qquad (3.224)$$

In the limit $|m| \to 1$, (3.224) is identical to PA. With this approximation in place, the exact series solution for perpendicular incidence can be summed to give (Sharma et al. 1997b):

$$T(\theta)_{HMA}^{TMWS} = -\frac{i\pi x\rho m^{1/2}\exp(i\rho/2)}{2[1+\tilde{r}^2\exp(4imx)]}\Big(\frac{J_1(x\omega(\cos\theta))}{x\omega(\cos\theta)}$$

$$+ i\tilde{r}\frac{J_1(x\omega(-\cos\theta))}{x\omega(-\cos\theta)}\exp(2imx)\Big), \qquad (3.225a)$$

and

$$T(\theta)_{HMA}^{TEWS} = \cos\theta\, T(\theta)_{HMA}^{TEWS}. \qquad (3.225b)$$

For $\tilde{r} \ll 1$, it further simplifies to,

$$T(\theta)_{HMA1}^{TMWS} = \frac{i\pi x\rho}{2} e^{i\rho/2} \frac{J_1(u)}{u}; \qquad T(\theta)_{HMA1}^{TMWS} = \cos\theta \; T(\theta)_{HMA1}^{TMWS}. \quad (3.226)$$

But for the factor $\exp(i\rho/2)$, expression (3.226) is identical to that in the PA and the modified RGA of Gordon (1985). For intensity, this difference is of no consequence.

The error in Q_{ext}^{HMA} for a homogeneous sphere is less than 5% for any x as long as $m \leq 1.10$. In contrast, the corresponding upper limit for Q_{ext}^{MPA} is $m = 1.06$. The errors in angular scattering have been examined by Sharma (1994). The range of $m - x$ values considered were $1.0 \leq x \leq 50$ and $1.05 \leq m \leq 1.50$. The HMA and the HMA1 agree with exact results very well for $\theta \leq 60$ deg. The minima appear to be slightly deeper in the HMA1. Otherwise, no significant difference from the HMA has been noted.

3.11 Evans and Fournier approximation

The empirical extinction efficiency factor obtained in the Evans and Fournier approximation (EFA) (Evans and Fournier 1990, 1994; Fournier and Evans 1991, 1996) for a homogeneous sphere is:

$$Q_{ext}^{EFA} = Q_{ext}^{R} \Big[1 + \Big(\frac{Q_{ext}^{R}}{T Q_{ext}^{ADA}} \Big)^{P} \Big]^{-1/P}, \quad (3.227)$$

where

$$Q_{ext}^{R} = \frac{24nn'}{z_1} x + \Big[\frac{4nn_i}{15} + \frac{20nn_i}{3z_2} + \frac{4.8nn_i[7(n^2+n_i^2)^2 + 4(n^2 - n_i^2 - 5)]}{z_1^2} \Big] x^3$$
$$+ \frac{8}{3} \Big[\frac{[(n^2+n_i^2)^2 + (n^2 - n_i^2 - 2)]^2 - 36n^2 n_i^2}{z_1^2} \Big] x^5, \quad (3.227a)$$

is the extinction efficiency to order x^4 in the RA and z_1 and z_2 are as given in (3.25) and (3.26). The parameters P and T are given by the relations:

$$P = A + \frac{\mu}{x}, \qquad T = 2 - exp(-x^{-2/3}). \quad (3.227b)$$

By extensive trial and error, the following expressions for A and μ were obtained:

$$A = \frac{1}{2} + \Big[n - 1 - \frac{2}{3}\sqrt{n_i} - \frac{n_i}{2} \Big] + \Big[n - 1 + \frac{2}{3}(\sqrt{n_i} - 5n_i) \Big]^2, \quad (3.227c)$$

and

$$\mu = \frac{3}{5} - \frac{3}{4}\sqrt{n-1} + 3(n-1)^4 + \frac{25}{6 + [5(n-1)/n_i]}. \tag{3.227d}$$

The formula for the extinction efficiency (3.227) is valid for spheroids too. Obviously, P and T are different and can be found in (Fournier and Evans 1991, Sharma and Somerford 2006).

Error contour charts have been plotted over the complex refractive index plane for extinction efficiency. The agreement with the exact results is reasonably good. For example, for $m \leq 1.62$ the error in the EFA does not exceed 20%.

3.12 Large particle approximations

3.12.1 Empirical formulas

For large particles with $|m| > 2$, asymptotic expressions for efficiency factors have been found to be (van de Hulst 1957, Kim et al. 1996),

$$Q_{ext} = 2.0 \tag{3.228}$$

and

$$Q_{sca} = \frac{1}{2}\left[F_1 + F_2\right], \tag{3.229}$$

where

$$F_i = \frac{8}{q_i^2}\left[q_i - \ln\left(1 + q_i + \frac{1}{2}q_i^2\right)\right], \quad i = 1, 2, \tag{3.230}$$

with

$$q_1 = \frac{1}{\sqrt{n_r n_i}} \quad \text{and} \quad q_2 = \frac{2}{q_1}. \tag{3.231}$$

An approximate formula given by Levine (1978) for the absorption efficiency factor of large spheres is:

$$Q_{abs} = 2(1-A)G(B), \tag{3.232}$$

where

$$G(B) = \frac{1}{2} + \frac{e^{-B}}{B} + \frac{e^{-B}-1}{B^2}, \tag{3.233}$$

with

$$B = \frac{2n_r\left[1 + 2/(n_r^2 + n_i^2)\right]}{1-A}, \tag{3.234a}$$

and

$$A = \left[\frac{(n_r-1)^2 + n_i^2}{(n_r^2+1)^2 + n_i^2}\right] + (A-1)e^{-B}. \tag{3.234b}$$

Equation (3.232) agrees with the exact results to within 2%. The refractive index range was $1.29 \leq n_r \leq 4.22$ and $0.0472 \leq n_i \leq 2.5259$.

3.12.2 Fraunhofer diffraction approximation

This approximation is most suited for near forward scattering of unpolarized waves by highly absorbing large obstacles. For a sphere, it gives the following expression for $i(\theta)$,

$$i(\theta)_{FDA} = 2x^4 \left[\frac{J_1(x\sin\theta)}{x\sin\theta}\right]^2 \times \frac{(1+\cos^2\theta)}{2}. \tag{3.235}$$

The first minimum of the function is governed by the term in the square bracket and is known to occur at,

$$\sin\theta = \frac{1.22\pi}{x}, \tag{3.235a}$$

As the size of the scatterer increases, the angular position of the first minimum decreases.

Jones (1977) has compared the FDA with Mie theory by generating error contour charts. The error contours give constant error trajectories in $x - m$ plane. The error was computed for the half width of the forward scattering lobe. For transparent particles, the error was less than 20% for $m > 1.3$ and $x \geq 20$. As expected, the error in absorbing particles is less. Less than 20% error occurs when $m > 1.2$ and $x \geq 6$. For nonabsorbing particles of very low refractive index ($m < 1.1$), significant error is caused by the interference of the transmitted and the diffracted wave giving rise to the anomalous diffraction.

The phase function for the scattering by an infinite cylinder at perpendicular incidence in the diffraction approximation is (Bohren and Huffman 1983),

$$\phi(\theta) = \frac{1}{8x}\left[\frac{1+\cos\theta}{\pi}\frac{x\sin(x\sin\theta)}{x\sin\theta}\right]^2. \tag{3.236a}$$

The phase function vanishes in the backward direction. Also,

$$d = \lambda n_{min}, \tag{3.236b}$$

where n_{min} is the n-th minimum and d is the diameter of the cylinder.

3.12.3 Geometrical optics approximation

Consider a ray hitting a sphere at an angle θ_i. A part of it is reflected and a part refracted. The process repeats itself an infinite number of times. If the angle between the N-th emerging ray and the direction of the incident ray is denoted by $\theta_N(\theta_i)$, the amplitude functions can then be written as (Glantschnig and Chen 1981):

$$S_j = \sum_{\theta_N(\theta_i)=\theta} S_j^{(N)}(x, m, \theta_i), \quad j = 1, 2, \tag{3.237}$$

where

$$S_j^{(N)}(x,m,\theta_i) = \sqrt{i_j^{(N)}(x,m,\theta_i)} \exp\left[i\sigma_N(x,m,\theta_i)\right],$$

$$i_j^{N)}(x,m\theta_i) = x^2\alpha_j^{(N)}(m,\theta_i)G^{(N)}(m,\theta_i),$$

$$\alpha_j^{(1)} = r_j^2; \quad \text{for} \quad N = 1,$$

$$\alpha_j^{(N)} = \left[(1-r_j^2)(-r_j)^{(N-2)}\right]^2; \quad \text{for} \quad N \geq 2,$$

$$r_1 = \frac{\cos\theta_i - m\ \cos\theta_r}{\cos\theta_i + m\ \cos\theta_r}, \quad r_2 = \frac{m\ \cos\theta_i - \cos\theta_r}{m\ \cos\theta_i + \cos\theta_r}, \tag{3.238}$$

and

$$G^{(N)}(m,\theta_i) = \frac{\sin\theta_i\cos\theta_i}{\sin\theta_N(\theta_i)} \times \left|\frac{\sqrt{m^2-\sin^2\theta_i}}{2[(N-1)\cos\theta_i - \sqrt{m^2-\sin^2\theta_i}]}\right|. \tag{3.239}$$

For forward scattering only two terms, $N = 1$ and $N = 2$ are important and simple analytic results have been obtained for this case.

3.12.4 Bohren and Nevitt approximation

The Bohren and Nevitt approximation (BNA) gives a simple analytic expression for Q_{abs} of a weakly absorbing sphere. Tracing ray paths, and summing all the contributions, it can be shown that for a homogeneous sphere (Bohren and Nevitt 1983, Bohren and Huffman 1983),

$$Q_{abs} = 2\int_0^{\pi/2} T(\theta_i,m)\frac{1-e^{-\alpha\xi}}{1-R(\theta_i,1/m)e^{-\alpha\xi}}\cos\theta_i\sin\theta_i\ d\theta_i, \tag{3.240}$$

where T and R are the transmittance and reflectance of the unpolarized light, θ_i is the angle of incidence, $\alpha = kn_i$ and ξ is the path length traversed by the ray from one crossing at the boundary to the next crossing:

$$\xi = \frac{2a}{n_r}\sqrt{n_r^2 - \sin^2\theta_i}. \tag{3.241}$$

By making a change of variable,

$$u = \frac{n_r^2 - \sin^2\theta_i}{n_r^2}, \tag{3.242}$$

equation (3.240) can be rewritten as,

$$Q_{abs} = n_r^2\int_{(n_r^2-1)/n_r^2)}^1 f(u)\left[1-\exp(-2\kappa\sqrt{u})\right] du, \tag{3.243}$$

with,

$$f(u) = \frac{T(u)}{1 - [1 - T(u)]\exp(-\tau\sqrt{u})}, \qquad (3.243a)$$

and $\kappa = 2xn_i$. Approximating $f(u) \approx 1$, the integration in (3.243) can be performed analytically to yield,

$$Q_{abs}^{BNA} = C_1 \left[\frac{1}{n_r^2} - \frac{2}{n_r^2} [e^{-\kappa\sqrt{n_r^2-1}/n_r} \left(1 + \frac{\kappa\sqrt{n_r^2-1}}{n_r} \right) - e^{-2\kappa}(1 + 2\kappa)] \right], \qquad (3.244)$$

where

$$C_1 = \frac{4n_r^3}{(n_r+1)^2 - (n_r-1)^2\exp(-2\kappa)}. \qquad (3.244a)$$

Equation (3.244) indeed yields adequate results over a wide range of sphere sizes. As κ increases, Q_{BNA}^{abs} approaches the limit $4n_r/(n_r+1)^2$, which is the transmittance of a plane surface for the normally incident wave.

For $\kappa \ll 1$, the right-hand side of (3.244) can be expanded in powers of κ. The expansion up to the third order in κ leads to

$$Q_{abs}^{BNA} = \frac{4}{3}\kappa n_r^2 (1 - \hat{b}^3). \qquad (3.245)$$

where

$$\hat{b} = \frac{(n_r^2 - 1)^{1/2}}{n_r}. \qquad (3.245a)$$

Similar formulas have been obtained by Pinchuk and Romanov (1977) and and Twomey and Bohren (1980) also.

The success of BNA prompted Flatau (1992) to propose a modified ADA (which he called AD theory) in which κ occurring in the ADA is replaced by κ_{new} defined as,

$$\kappa_{new} = n_r^2(1 - b^3)\kappa. \qquad (3.246)$$

The absorption efficiency factor so obtained is found to be in close agreement with the BNA, for all values of κ except near unity.

A comparison of the absorption efficiency factor from (3.244) with exact results for a water droplet of $m = 1.304 + i0.001082$ in the size range $0 \leq a \leq 100\ \mu m$ at $\lambda = 2\ \mu m$ shows that the BNA is adequate over the entire size range (Bohren and Nevitt 1983). Flatau (1992) has compared Q_{abs}^{BNA}, Q_{abs}^{ADA} and the corrected Q_{abs}^{ADA} (κ given by 3.246)) against κ for fixed n_r. It was found that whereas the standard ADA differs from the BNA in most of the κ region, the corrected ADA shows excellent agreement with the BNA up to $\kappa \sim 0.5$.

3.12.5 Nussenzweig and Wiscombe approximation

This is an asymptotic approximation in the size parameter ($x \to \infty$). Real and imaginary parts of $S(0)$ for a sphere in this approximation can be expressed as (Nussenzweig and Wiscombe 1980):

$$\frac{4}{x^2}\,\mathrm{Re}S(0)^{NWA} = Q_{ext}^{NWA} = 2 + 1.9923861x^{-3/2} - 0.7153537x^{-4/3}$$

$$-0.3320643(\frac{3^{1/2}}{2})\frac{(m^2+1)(2m^4-6m^2+3)}{(m^2-1)^{3/2}}x^{-5/3} - \frac{16m^2\sin\rho}{(m+1)^2\rho}$$

$$-\frac{16m(3m-m^2-1)\cos\rho}{\rho^2} - \frac{8}{x}\sum_{j=1}^{\infty}\frac{(m-1)}{(2j+1-m)}$$

$$\left[\frac{m-1}{m+1}\right]^{2j}\sin 2(m_1+2jm)x + O(1/x^2) + ripple, \qquad (3.247a)$$

and

$$\frac{4}{x^2}\,\mathrm{Im}S(0)^{NWA} = -3^{1/2}\left(1.9923861x^{-3/2} - 0.7153537x^{-4/3}\right) + \frac{2(m^2+1)}{x(m^2-1)^{1/2}}$$

$$-0.3320643\left(\frac{3^{1/2}}{2}\right)\frac{(m^2+1)(2m^4-6m^2+3)}{(m^2-1)^{3/2}}x^{-5/3} - \frac{16m^2\cos\rho}{(m+1)^2\rho}$$

$$-\frac{16m(3m-m^2-1)\cos\rho}{\rho^2} - \frac{8}{x}\sum_{j=1}^{\infty}\frac{(m-1)}{(2j+1-m)}$$

$$\left[\frac{m-1}{m+1}\right]^{2j}\sin 2(m_1+2jm)x + O(1/x^2) + ripple, \qquad (3.247b)$$

Numerical comparison of Q_{ext}^{NWA} and $i(0)^{NWA}$ with exact quantities as well as with the corresponding quantities in the PA, the MPA, the SPA and the MSPA for nonabsorbing spheres has been done by Roy and Sharma (1996). The last two terms on the right-hand side of (3.251a) and (3.251b) have been ignored in numerical computations. It has been observed that (i) if x is grater than a certain value (say $x_>$), NWA is better than any other approximation considered. (ii) The value of $x_>$ is dependent on $|m-1|$. It decreases as $|m-1|$ increases. For example, we observe that for $m = 1.02$, $x_> = 374$; $m = 1.06$, $x_> = 82$ and for $m = 1.16$, $x_> = 26$.

3.13 Other large size parameter approximations

Shifrin and Tonna (1992) have arrived at the following approximation for weakly refracting small particles:

$$Q_{abs}^{STA} = \left[1 - e^{-\frac{4}{3}\kappa n^2(1-\hat{b}^3)}\right]. \tag{3.248}$$

If the exponential term in (3.247) is expanded for small absorption, the leading term in the expansion coincides with (3.245).

Kokhanovsky and Zege (1997) have shown that if $(n_i < 0.1)$, a simple formula for the absorption efficiency factor can be written as:

$$Q_{abs}^{KZA} = \bar{T}\Bigg(1 - \frac{n_r^2}{8n_i^2x^2}\left[e^{-4n_ix\hat{b}}(1+4n_ix\hat{b}) - e^{-4n_ix}(1+4n_ix)\right]$$

$$-S(n)\left[1 - e^{-4n_ix\hat{b}}\right]^2\Bigg), \tag{3.249}$$

where the subscript KZA stands for Kokhanovsky and Zege approximation,

$$\bar{T} = 1 + (n_r - 1)\left(1 - e^{-1/(\bar{t}\rho)}\right), \tag{3.249a}$$

$$\bar{t} = \left[21.2 - 20.1 \ \log \ n_i + 11.1(\log \ n_i)^2 + (\log \ n_i)^3\right]^{-1}, \tag{3.249b}$$

and

$$S(n) = \left[\frac{8n_r^2(n_r^4+1)\ln \ n_r}{(n_r^4-1)^2(n_r^2+1)} - \frac{n_r^2(n_r^2-1)^2}{(n_r^2+1)^3}\ln\frac{n_r+1)}{(n_r-1}\right.$$

$$\left.+\frac{3n_r^7 - 7n_r^6 - 13n_r^5 - 9n_r^4 + 7n_r^3 - 3n_r^2 - n_r - 1}{3(n_r^4-1)(n_r^2+1)(n_r+1)}\right]. \tag{3.249c}$$

As usual $\rho = 2x(n_r - 1)$.

Another expression, obtained directly in the framework of complex angular momentum theory, for weakly absorbing soft particles $(n_r \le 1.2)$ is (Kokhanovsky and Zege 1995)

$$Q_{abs}^{K} = Q_{abs}^{BNA} + Q_{abs}^{edge}, \tag{3.250}$$

where

$$Q_{abs}^{edge} = 4n_rn_ix\left[\cos^{-1}\left(\frac{1}{n_r}\right) - \frac{1}{n_r^2}\sqrt{n_r^2-1}\right]. \tag{3.250a}$$

Numerical evaluation shows that the error in (3.248) does not exceed 10% if $x \geq 10.0$ and the refractive index range $1.2 \leq n_r \leq 1.55$. Kokhanovsky and Zege (1997) compared the KZA, the STA, and the BNA against the Mie absorption efficiency factors. Comparisons were made for $n = 1.34$, $20 \leq x \leq 200$, $n_i = 10^{-3}$ for $2n_i x < 1$ and $n_i = 10^{-2}$ for $2n_i x > 1$. The STA was found to be best among the three approximations.

An approximation for large semi-transparent particles, suggested by Dombrovsky (2002) is:

$$Q_{abs} = \frac{n_r}{(n_r + 1)^2} \left[1 + e^{-4n_i x}\right]. \tag{3.251}$$

The approximation yields good results for large x.

If Re $S(0) \gg$ Im $S(0)$, the scattered intensity may be approximated as (Penndorf 1962),

$$i(0) \approx [\text{Re } S(0)]^2 \approx \frac{x^4}{4} \left(\frac{Q_{ext}}{2}\right)^2. \tag{3.252}$$

Further, since

$$i_{FDA}(0) = \frac{x^4}{4} \tag{3.253}$$

is a good approximation for a large sphere, (3.252) may be written as,

$$i_{PSPA}(\theta) = i_{FDA}(\theta) \left(\frac{Q_{ext}}{2}\right)^2, \tag{3.254}$$

for near forward scattering (Shifrin and Punina 1968). The subscript $PSPA$ stands for Penndorf–Shifrin–Punina approximation.

A detailed analysis of errors in $i(0)_{PSPA}$, led Fymat and Mease (1981) to propose the following approximation for $i(\theta)$:

$$i(\theta)_{FMA} = i(\theta)_{PSPA} f_1 f_2, \tag{3.255}$$

where

$$f_1^{-1} = 1 - J_0^2[(m-1)x], \quad \text{for } (m-1)x \leq \alpha \text{ or } \beta \leq (m-1) \leq \gamma \tag{3.255a}$$

or 1 otherwise and

$$f_2^{-1} = 1 - J_0^2[2(m-1)x] \text{ for } 2(m-1)x \leq \delta, \tag{3.255b}$$

or 1 otherwise. The constants α, β, γ, δ are m dependent constants. For $m = 1.33$, these are found to be $\alpha = 3.63$, $\beta = 5.52$, $\gamma = 6.6$ and $\delta = 2.40$. This formula gives reasonable agreement with the exact results for near forward scattering by nonabsorbing spheres.

3.14 Composite particles

3.14.1 Effective medium theories

Effective medium theories (EMTs) aim at replacing a heterogeneous scatterer by an equivalent homogeneous one by identifying an effective refractive index or dielectric constant. Let f_1 and f_2 be the volume fractions of two components with dielectric constants ε_1, and ε_2, respectively. If the inclusion dielectric constant is ε_1 and the host dielectric constant is ε_2, then the most commonplace mixing rules are:

1. Lorentz–Lorenz (Lorentz 1880, Lorenz 1880):

$$f_1\frac{\varepsilon_1 - 1}{\varepsilon_1 + 2} = \frac{\varepsilon_{eq} - 1}{\varepsilon_{eq} + 2}; \quad \varepsilon_2 = 1, \tag{3.256a}$$

2. Maxwell–Garnett (MS) (1904):

$$f_1\frac{\varepsilon_1 - \varepsilon_2}{\varepsilon_1 + 2\varepsilon_2} = \frac{\varepsilon_{eq} - \varepsilon_2}{\varepsilon_{eq} + 2\varepsilon_2}, \tag{3.256b}$$

3. Bruggeman (1935):

$$f_1\frac{\varepsilon_1 - \varepsilon_{eq}}{\varepsilon_1 + 2\varepsilon_{eq}} = \frac{\varepsilon_2 - \varepsilon_{eq}}{\varepsilon_{eq} + 2\varepsilon_2}. \tag{3.256c}$$

Numerical tests show that if $x \leq 0.1$, the MG and Bruggeman rule provide accurate values for the extinction and scattering efficiency factors. The errors are within 1%. For absorption efficiency however, these increase to 10-15% (Chylek et al. 2000).

Some other simple mixing rules are:

$$\varepsilon_{eq} = \varepsilon_1\left[1 + \frac{3(1-f)(\varepsilon_2 - \varepsilon_1)/(\varepsilon_1 + \varepsilon_2)}{1 - (1-f)(\varepsilon_2 - \varepsilon_1)/(\varepsilon_1 + \varepsilon_2)}\right]; \quad \text{inverse MG}, \tag{3.257a}$$

$$\varepsilon_{eq}^{1/3} = f\varepsilon_1^{1/3} + (1-f)\varepsilon_2^{1/3}; \quad \text{Looyenga (1965)}, \tag{3.257b}$$

$$\varepsilon_{eq}^{1/2} = f\varepsilon_1^{1/2} + (1-f)\varepsilon_2^{1/2}; \quad \text{Birchak (1974)}, \tag{3.257c}$$

$$\log\varepsilon_{eq} = f\log\varepsilon_1 + (1-f)\log\varepsilon_2; \quad \text{Lichtenecker (1926)}, \tag{3.257d}$$

and

$$\varepsilon_{eq} = \left[f\varepsilon_1^{\alpha} + (1-f)\varepsilon_2^{\alpha}\right]^{1/\alpha}; \quad \text{Landau and Lifshitz (1960)}, \tag{3.257e}$$

where α is a constant in the range $(-1 \leq \alpha \leq 1)$. Lumme and Rathola (1994) have given a formula which reduces to the MG formula in one limit and Bruggeman theory in another limit.

The EMAs can be generalized for the case of an N component mixture by writing (Aspnes 1982):

$$\frac{\varepsilon_{eq} - \varepsilon_0}{\varepsilon_{eq} + 2\varepsilon_0} = f_1 \frac{\varepsilon_1 - \varepsilon_0}{\varepsilon_1 + 2\varepsilon_0} + f_2 \frac{\varepsilon_2 - \varepsilon_0}{\varepsilon_2 + 2\varepsilon_0} + ..., \tag{3.258}$$

For $\varepsilon_0 = 1$ it reduces to the Lorentz–Lorenz rule (3.256a). The MG rule (3.256b) is obtained by assuming one of the components to be the host medium ($\varepsilon_0 = \varepsilon_2$). The Bruggeman rule (3.256c) is obtained if $\varepsilon_0 = \varepsilon_{eq}$.

A basic requirement for the validity of the EMAs is that the typical dimension of the inclusions must be much smaller than the wavelength of the incident radiation. This is because the electrostatic approximation is the basic assumption in all the EMAs. The validity of the electrostatic approximation requires $x \ll 1$. In practice however, the EMAs can yield good results even for inclusion sizes as large as $x \sim 0.5$ (Niklasson and Granqvist 1984).

A modification to the MG rule, which allows for larger size inclusions, has been given by Lakhtakia and Vikram (1993):

$$m_{eq} = m_1 \frac{(1 + (2\alpha f_v)/3)^{1/2}}{(1 - (2\alpha f_v)/3)^{1/2}}, \tag{3.259}$$

where

$$\alpha = \frac{(m_2/m_1)^2 - 1}{1 - [(m_2/m_1)^2 - 1]\,[(2/3)(1 - ikam_1)\exp(ikam_1) - 1]}. \tag{3.259a}$$

Subscripts 1 and 2 in the two equations above refer to the surrounding medium and particles, respectively. The method is valid if $|kam_j| \leq \pi/5$, where $j = 1, 2$ and $0 \leq f_v \leq 0.2$.

Voshchinnikov et al. (2007) have studied extinction efficiency factors and scattering properties of heterogeneous particles using various EMAs. The approximate results obtained by applying EMAs were compared with theexact results obtained using the discrete dipole approximation (DDA). It was concluded that the Bruggeman prescription gave the least deviation from the exact results. The errors in the extinction factor did not exceed 5% for particle porosity $\mathcal{P} = 0 - 0.9$ and size parameter $x_p = 2\pi r_p/\lambda \leq 0.25$. The porosity $\mathcal{P} = 0$ corresponds to a compact homogeneous sphere. The relation,

$$x_p = \frac{x_c}{(1 - \mathcal{P})} = \frac{x_c}{(V_{solid}/V_{total})^{1/3}}, \tag{3.260}$$

gives the size parameter of the porous particle x_p in terms of the compact size parameter x_c.

For highly absorbing particles, extended EMAs (EEMAs) have been developed which are valid if,

$$2\pi d \frac{\text{Im } m_{eq}}{\lambda} \ll 1, \tag{3.261}$$

where d is the inclusion size. For an inclusion size parameter up to $x = 2$, the MG and Bruggeman rules predict extinction and scattering efficiency factors with errors between 10-15% for water inclusions in an acrylic matrix and black carbon in water droplets (Chylek et al. 2000). In the same size parameter range, the errors in the EEMA are between 3-5%.

Under certain circumstances it is possible to treat the clusters of spheres as a compact sphere. If a_1 denotes the radius of the primary particle of an N particle aggregate, and if $a_{N,eq}$ is the radius of the equivalent sphere, then the volume fraction of the matter in the equivalent sphere is,

$$\bar{\phi}_a = \frac{N\ a_1^3}{a_{N,eq}^3}. \tag{3.262}$$

If the aggregate is approximated as a sphere containing all the matter ($\bar{\phi}_a = 1$) the approximation is called a compact sphere approximation (CSA). Jacquier and Gruy (2010) show that this method can be valid for aggregates of high compactness. It has been found to overestimate the scattering cross section for $x < 7.0$ whatever the configuration of the aggregate. Drolen and Tien (1987) and Dobbins and Megaridis (1991) have also approximated agglomerates of spherical particles as a solid sphere using MG theory.

Grenfell and Warren (1999) and Grenfell et al. (2005) modeled ice crystals as a collection of spheres to compute the optical properties of ice crystals conserving a volume to area ratio. The validity of this approximation with respect to various ice crystal shapes has been found to be good. However, its use for more complex-shaped fractal soot aggregates by Li et al. (2010) shows substantial errors. Equivalent volume spheres, equivalent surface spheres and equivalent radius spheres were also included in this comparison.

3.14.2 Effective refractive index method

Jacquier and Gruy (2010) have approximated an aggregate of spheres by an equivalent sphere by imposing the following conditions: (i) the effective refractive index m_{eq} is defined using the MG theory as,

$$\frac{m_{eff}^2 - 1}{m_{eff}^2 + 2} = \bar{\phi}_a \frac{m^2 - 1}{m^2 + 2}, \tag{3.263}$$

and (ii) the projected area equivalent sphere is defined by the relation

$$\pi a_{N,eq}^2 = \langle S_p \rangle, \tag{3.264}$$

where $\langle S_p \rangle$ is the average projected area according to all aggregate orientations. The ADA is used for further treatment.

For a large sphere with highly absorbing randomly distributed spherical inclusions in it, an approximate formula based on a geometrical optics approximation was developed by Sharma and Jones (2000). The model assumed that the rays propagated undisturbed in the large sphere but were completely absorbed on hitting an inclusion. The real part of the equivalent refractive index of the large sphere was taken to be the same as the host medium, but its imaginary part was taken to be governed by the relation,

$$n_{i,eq} = \frac{3f_v}{8x_i} + (1 - f_v)n_i, \tag{3.265}$$

where n_i is the imaginary part of the refractive index of inclusions, x_i is the size parameter of the inclusions and f_v is the volume fraction of the inclusions. The Mie computations for the resulting equivalent sphere were compared against exact results from the ray tracing Monte Carlo method. This study demonstrated that (3.265) is indeed a good approximation for predicting the absorption efficiency, asymmetry parameter and albedo.

In a subsequent study, Sharma and Jones (2003) extended the approximation (3.265) to an absorbing host medium as well. The resulting empirical term for the imaginary part of the equivalent refractive index is,

$$n_{i,eq} = \frac{3f_v}{8x_i} + \frac{1}{25f_v(1 + x_i)}(1 - f_v)m_{hi}, \tag{3.266}$$

where x_i is the size parameter of the inclusions, and m_{hi} is the imaginary part of the host medium refractive index. It was concluded from numerical comparisons that the equivalent refractive index method was accurate when the volume concentration of inclusions is small and the difference between the two refractive indices is not large.

Chapter 4

Scattering by an assembly of particles

A simple graphical description of scattering of electromagnetic radiation by a particulate medium is exhibited as an infinite series of ladder diagrams in Figure 4.1. The first ladder diagram represents no interaction. These photons are called ballistic photons. The second graph represents single scattering or the scattering of order 1. Other graphs represent higher order scatterings or multiple scatterings. The figure shows only upto third order scattering. The photons that have undergone one or more than one scattering but are still travelling close to the forward direction are called quasi-ballistic photons or snake photons. The remaining photons are diffuse photons. Our focus in this chapter is limited mainly to the single scattering perspective.

Since the single scattering setting is a smooth extension of single particle scattering, analytic formulas for observables such as the extinction spectrum, phase function, asymmetry parameter, etc., have been obtained for a variety of particle shapes and in a variety of size distributions. Physical parameters, such as refractive index and the size distribution of scatterers, can be extracted from experimental measurements using such analytic formulas. Occasionally, the single scattering formulas can be utilized even in a dense collection. This can be achieved by diluting the sample under investigation. But, care has to be exercised, so that the process of dilution does not alter the properties of the original ensemble.

Single scattering analytic relationships are important from the multiple scattering point of view too. Single scattering quantities, such as phase function and scattering and absorption coefficients, are required as input for solving multiple scattering problems. Simple analytic expressions for these input parameters often result in a reduction of complexity of the multiple scattering problem. Thus, the suitability of a large number of phase functions has been studied for various particle types. Related problems, like the relationship of a single scattering phase function with higher order phase functions for quasi ballistic scattering photons and reflectance at an arbitrary source detector separation, are also included to some extent. This aspect is important from the

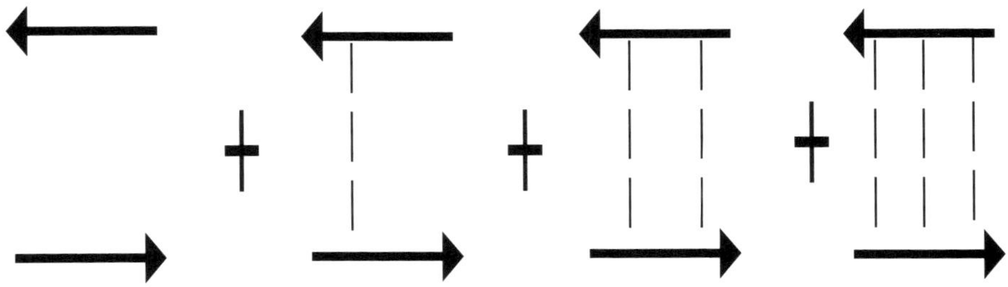

FIGURE 4.1: A typical ladder diagram depicting various orders of scattering.

point of view of the inverse problem of single scattering phase function quantification from the measured effective phase function or from the measured wave reflectance. Analytic solutions of the radiative transfer equation itself are not addressed here. We refer the readers to well-known texts on multiple scattering problems including those by Chandrasekhar 1960, Ishimaru 1999, Mishchenko et al. 2006, Dombrovsky and Basillis 2010, and review articles by Kokhanovsky (2002) and Borovoi (2006).

4.1 Single scattering by $\mathcal{N}$ independent particles

Consider the scattering of electromagnetic radiation by two independent particles separated by a distance $\mathbf{r}$. The detector is at a distance $\mathbf{R}$ from the origin of the coordinate system such that $\mathbf{r} \ll |\mathbf{R}|$. If the scattered fields by particles 1 and 2 at the detector are denoted by

$$\mathbf{E}_1(\mathbf{R},t) = E_1\ e^{i(\mathbf{k}_s.\mathbf{R}-\omega t+\phi_1)}, \tag{4.1}$$

$$\mathbf{E}_2(\mathbf{R},t) = E_2\ e^{i(\mathbf{k}_s.\mathbf{R}-\omega t+\phi_2)}, \tag{4.2}$$

respectively, the resultant field is,

$$\mathbf{E}_{tot} = \mathbf{E}_1(\mathbf{R},t) + \mathbf{E}_2(\mathbf{R},t). \tag{4.3}$$

The intensity at the detector is then (see eq. (1.30)),

$$I_{tot} = (\frac{1}{2}\varepsilon c)\left[E_1^2 + E_2^2 + 2E_1E_2\cos\delta\phi\right], \tag{4.4}$$

where ε is the dielectric constant of the host medium and

$$\delta\phi = (\phi_2 - \phi_1) = (\mathbf{k}_i - \mathbf{k}_s), \mathbf{r}. \tag{4.5}$$

If the two scatterers are identical ($E_1^2 = E_2^2 = E_0^2$),

$$I_{tot} = (\frac{1}{2}\varepsilon c E_0^2)\ 2\left[1 + \cos\delta\phi\right]. \tag{4.6}$$

Thus, depending on the value of $\delta\phi$, the scattered intensity can be anything between 0 to 4 times the intensity scattered by a single particle.

The result obtained for two scatterers can be generalized to scattering by an ensemble of identical particles. If $\mathcal{N}$ denotes number of particles per unit volume in the ensemble, the scattered intensity by particles in this unit volume can be written as,

$$I_{tot,\mathcal{N}} = \frac{1}{2}\varepsilon c E_0^2 \sum_{j=1}^{\mathcal{N}}\sum_{k=1}^{\mathcal{N}} \cos(\phi_j - \phi_k), \tag{4.7}$$

and the mean scattered intensity becomes,

$$\langle I_{tot,\mathcal{N}} \rangle = \frac{1}{2}\epsilon c E_0^2 \sum_{j=1}^{\mathcal{N}} \sum_{k=1}^{\mathcal{N}} \langle \cos(\phi_j - \phi_k) \rangle . \tag{4.8}$$

If the particles are moving independently of each other, the phases ϕ_j and ϕ_k will change all the time randomly. Hence, $\langle \cos(\phi_j - \phi_k) \rangle = 0$, if $j \neq k$. For $j = k$, $\langle \cos(\phi_j - \phi_k) \rangle = 1$ and

$$\langle I_{tot,\mathcal{N}} \rangle = (\mathcal{N}) \frac{1}{2} \varepsilon c E_0^2 . \tag{4.9}$$

That is, the intensity of the wave scattered by an assembly of $\mathcal{N}$ randomly distributed scatterers is $\mathcal{N}$ times the intensity scattered by a single scatterer.

It follows that the differential scattering cross-sections also add. The scattering coefficient may thus be defined as,

$$\mu_{sca} = \mathcal{N} C_{sca} . \tag{4.10}$$

This is nothing but the average number of scatterings per unit length. In a similar manner, the absorption coefficient may be defined as,

$$\mu_{abs} = \mathcal{N} C_{abs} , \tag{4.11}$$

and the sum of (4.10) and (4.11) gives,

$$\mu_{ext} = \mu_{sca} + \mu_{abs} , \tag{4.12}$$

as the extinction coefficient. Often, the letters K and $b(\lambda)$ are also used in place of μ.

If the constituents of the ensemble differ in size, a particle size distribution $n(a)$ may be defined, such that, there are $n(a)da$ particles per unit volume having radii in the range a to $a + da$. Then

$$\mathcal{N} = \int_0^\infty n(a) da , \tag{4.13}$$

is the number of particles per unit volume. The scattering coefficient may then be defined as

$$\mu_{sca} = \int_0^\infty n(a) C_{sca} da . \tag{4.14}$$

Differing particle types may be included by writing,

$$\mu_{sca} = \sum_j \int_0^\infty n_j(a) C_{sca,j} da , \tag{4.15}$$

where n_j are the number of particles of type j per unit volume. Expressions analogous to (4.14) and (4.15) hold for the absorption and extinction coefficients too.

The extinction coefficient is related to turbidity via the Beer–Lambert law which expresses the transmitted intensity through a slab of width L as,

$$I = I_{inc}\exp\Big(-\int_0^L \mu_{ext}dx\Big) = I_{inc}\exp(-\tau), \tag{4.16}$$

where τ is the optical depth or turbidity of the medium. For a homogeneous medium, μ_{ext} is constant. Thus, $\tau = \mu_{ext}\ L$. An underlying requirement in the derivation of the Beer–Lambert law is that the scattering be negligible. Nevertheless, in practice it has been shown to be applicable even when appreciable scattering is present (Swanson et al. 1999).

The inverse of the extinction coefficient, $l = 1/\mu_{ext}$ is the optical mean free path length. It represents the average distance travelled by the wave between two scatterings. Thus, the condition for the absence of multiple scattering may be expressed as $L \ll l$ or $\tau\ < 1$. In practice however, significant contribution from multiple scattering may appear down to $\tau \sim\ 0.1$.

4.2 Multiple scattering

In a denser medium, the incident wave may undergo multiple scatterings before it finally exits the medium. An important parameter in the description of this process, is the scattering mean free path . It is defined as,

$$l_s = \frac{1}{\mu_{sca}}. \tag{4.17}$$

This is the average distance which the wave travels before encountering another scattering. One also defines a quantity called reduced mean free path length. This is a lumped parameter consisting of l_s and the anisotropy of the phase function is about $\theta = 90\,\text{deg}$. The following equation,

$$l_s' = \frac{l_s}{1-g} = \frac{1}{\mu_{sca}(1-g)} = \frac{1}{\mu_{sca}'}, \tag{4.18}$$

defines the relationship between l_s and l_s'. The reduced scattering coefficient μ_{sca}' represents the probability of an isotropic scattering event per unit length. The relation $\mu_{sca}' = \mu_{sca}(1-g)$ associates $1/(1-g)$ anisotropic steps with one isotropic step. For isotropic scattering $g = 0$ and $l_s' = l_s$, otherwise $l_s' > l_s$. For g close to 1, $l_s' \gg l_s$.

In general, the scattering may be accompanied by some absorption too. The mean free path length for absorption is then defined as,

$$l_a = \sqrt{\frac{1}{3} l_i l_s} = \frac{1}{\mu_{abs}}, \tag{4.19}$$

where l_i, the inelastic mean free path length, is the distance over which its intensity reduces by a factor of $1/e$ due to absorption.

4.3 Diffusion approximation

When scattering is the dominant process, it disperses the photons randomly. At very high concentrations and for high relative refractive index, diffusion becomes the dominant process for transport of photons. The energy transport process can then be well understood in terms of diffusion approximation. Mathematically, the approximation is applicable when

$$\lambda \ll l_s' \ll L \ll l_{abs}. \tag{4.20}$$

The condition implies a high concentration of scatterers. In terms of asymmetry parameter and albedo, the diffusion theory provides good approximation when

$$g \leq 0.1; \qquad \omega \rightarrow 1. \tag{4.21}$$

The first condition implies isotropic scattering and the second condition indicates a scattering dominant medium.

4.4 Radiative transfer equation

Radiative transfer equation provides a mathematical description of electromagnetic wave propagation in a cloud of scatterers. Each scatterer is assumed to be scattering independently. The radiative transfer equation can be derived on the basis of implementing the energy balance condition between incoming, scattered, absorbed, outgoing and emitted photons within a small volume element. This volume element is chosen such that only the single scattering occurs within this small volume. Choosing such a volume element is always possible. If the intensity is the only quantity of interest, electromagnetic scattering may be treated in the scalar approximation. Assuming that there are no sources of radiation inside the ensemble, the stationary radiative transfer equation may be expressed as,

$$\frac{\partial I(\mathbf{r},\mathbf{s})}{\partial s} = -\mu_{ext} I(\mathbf{r},\mathbf{s}) + \frac{\mu_{sca}}{4\pi} \cdot \int_{4\pi} I(\mathbf{r},\mathbf{s}')p(\mathbf{s},\mathbf{s}')d\Omega', \tag{4.22}$$

where $I(\mathbf{r}, \hat{s})$ is the radiation intensity at a point $\mathbf{r}$ in the direction $\mathbf{s}$, $d\Omega'$ is the unit solid angle in the direction $\mathbf{s}'$, and $p(\mathbf{s}, \mathbf{s}')$ is the scattering phase function of the particles in the infinitesimal volume element.

It is evident from (4.22) that the solution of the radiative transfer equation requires the scattering coefficient, absorption coefficient and phase function as input. If the single particle scattering and absorption efficiencies are known, μ_{abs} and μ_{sca} can be easily obtained by multiplying single scattering cross-sections with particle density when the scatterers are identical. In the event the scatterers display a size distribution, the scattering coefficient may be expressed as,

$$\mu_{sca} = \int_{d_0}^{d_m} \frac{\eta(d_i)}{v_i} C_{sca}(d_i)\; dd_i, \tag{4.23}$$

where v_i is the volume of a particle of size d_i and $\eta(d_i)$ is the volume fraction of particles with diameter d_i. The minimum and maximum in the population are d_0 and d_m, respectively. An analogous expression holds for μ_{abs}.

The phase function for a collection of polydisperse population, is defined as,

$$p(m,k,d_0,d_m,\theta,\varphi) = \frac{\int_{d_0}^{d_m} \phi(m,k,d_i,\theta,\varphi)\sigma_{sca}(m,k,d_i,\theta,\varphi)\eta(d_i)\; dd_i}{\int_{d_0}^{d_m} \sigma_{sca}(m,k,d_i,\theta,\varphi)\eta(d_i)ddi}, \tag{4.24}$$

and the normalization being followed here is,

$$\int_0^{2\pi}\int_0^{\pi} p(m,k,d_0,d_m,\theta,\varphi)\sin\theta d\theta d\varphi = 1. \tag{4.25}$$

In a discrete particle model, one needs to replace integrals by summations. Notice that while the phase function is independent of particle density, μ_{sca} and μ_{abs} are dependent on the particle concentration. A widely used alternative normalization condition is:

$$\int_0^{2\pi}\int_0^{\pi} p_{alt}(m,k,d_0,d_m,\theta,\varphi)\sin\theta d\theta d\varphi = 4\pi. \tag{4.25a}$$

Thus, $p = p_{alt}/4\pi$ is the conversion rule.

The azimuthal dependence of the scattering phase function is often removed, which is possible under an assumption of spherical symmetry. In that case,

$$2\pi \int_0^{\pi} p(m,k,d_0,d_m,\theta,\varphi)\sin\theta d\theta = 1. \tag{4.26}$$

is the relevant normalization condition for the phase function.

As the scatterer concentration increases, correlated scattering begins affecting whole cross section. The total scattering cross section decreases. The magnitude of this decrease is dependent on the size and the volume concentration of scatterers. For Rayleigh particles, the decrease in the total scattering cross section becomes significant when the volume concentration becomes greater than 0.01. For larger scatterers, this effect comes into play at a higher volume fraction of about 10%. The effect of reduction in the total cross section at higher concentrations can be accounted for by replacing the volume fraction $\eta(d_i)$ by an effective volume fraction given by (Twersky 1975),

$$\eta_{eff}(d_i) = \frac{(1-\eta(d_i))^4}{(1+2\eta(d_i))^2}\eta(d_i). \tag{4.27}$$

That is to say, (4.23) can be used at even higher concentrations by using (4.27). The relationship (4.27) was later modified by Bascom and Cobbold (1995) to includes media composed of different scatterer shapes:

$$\eta_{eff}(d_i) = \frac{[1-\eta(d_i)]^{p+1}}{[1+(p-1)\eta(d_i)]^{p-1}}\eta(d_i). \tag{4.28}$$

For spherical particles $p = 3$. In that case (4.28) reduces to (4.27).

Finally, the volume averaged backscattering coefficient is defined as,

$$\mu_{back} = \int_{d_0}^{d_m} \frac{\eta(d_i)}{v_i} C_{sca}(d_i) p(m, k, \theta = \pi, d_i) \, dd_i. \tag{4.29}$$

This has been used to characterize the reflectivity of a sample (Marchesini et al. 1989).

4.5 Phase function

There are two main approaches for constructing analytic phase functions. (i) When it is feasible to measure a phase function experimentally, an analytic form for it may be created by parameterizing the observed phase function. Such phase functions then provide significant help in solving the difficult multiple scattering problem. But, a drawback of phase functions formulated in this way is that they do not deal with the underlying physics directly. Thus, they do not relate to microphysical characteristics of the particles categorically. In general, quantities such as particle size distribution and refractive index cannot be extracted from such empirical relations. However, there are occasional exceptions, where it has been possible to relate these empirical parameters to microphysical characteristics of particles (Roy and Sharma 2008, Sahu and Shanmugam 2015). (ii) In the second approach, the analytic phase function is constructed using approximate single particle phase functions and cross-sections as input in (4.24). This second approach is more suitable for obtaining formulas useful for extracting information on particle characteristics. In place of approximate phase functions, exact phase functions may also be used in (4.24), whenever available. But, then the simplicity of the phase function is lost.

4.5.1 The Henyey–Greenstein phase function (HGPF)

In radiative transfer calculations, this is one of the most frequently used phase functions (Henyey and Greenstein 1941). Historically, this empirical formula originated in connection with astrophysics problems but was later adapted in a variety of other settings. It is a one-parameter phase function and may be expressed as,

$$p(\theta)_{HGPF} = \frac{1}{4\pi}\frac{1-g^2}{(1+g^2-2g\cos\theta)^{3/2}}, \tag{4.30}$$

or as a series in terms of the associated Legendre polynomials,

$$p(\theta)_{HGPF} = \frac{1}{4\pi}\sum_{l=0}^{\infty}(2l+1)g^l P_n(\cos\theta). \tag{4.31}$$

Notice that parameter g is,

$$g = 2\pi\int_{-\pi}^{\pi} p(\theta)_{HGPF}\cos\theta\sin\theta d\theta. \tag{4.32}$$

Clearly, parameter g is a measure of the asymmetry of the phase function around $\theta = 90\,\text{deg}$. For this reason, g is usually set equal to the asymmetry parameter of the phase function to be parametrized. Refined methods for determining the parameter g have been suggested (Kamiuto 1987, Pomraning 1988), but rarely used.

It is amusing to note that despite its huge popularity, no unambiguous criterion regarding its validity has been established yet. Nevertheless, this drawback has not impeded its use. It remains one of the most prevalent phase functions even after more than 7 decades in service. The foremost reason for its long shelf life appears to be its simple structure. In ray-tracing multiple scattering simulations, the HGPF allows determination of the random direction of a scattered photon through an analytic expression. Thus, if ξ is a random number, uniformly distributed over the interval 0 to 1, the random scattering angle can be shown to be given by the expression,

$$\cos\theta_{HGPF} = \frac{1}{2g}\left(1 + g^2 - \left[\frac{1 - g^2}{1 - g + 2g\xi}\right]^2\right), \tag{4.33}$$

if $g \neq 0$ and $\cos\theta = 2\xi - 1$ if $g = 0$. This makes the procedure of ray tracing simulation very efficient and fast.

The accuracy of the HGPF has been evaluated by numerical comparisons *vis a vis* exact phase functions. It has been noted that the HGPF is a monotonic decreasing function of scattering angle. Hence, it can neither reproduce the oscillations present in a single particle phase function, nor can it mimick the backscattering rise seen in most realistic phase functions. At best, one may expect the HGPF to reproduce a forward scattering lobe and/or average shape of the phase function. For near-forward scattering, numerical comparisons show that it underestimates the scattering (Sharma et al. 1998). Figure 4.2 shows this for $x = 1$ ($g = 0.2$) and $m = 1.5$. Similar conclusions have been drawn for scattering by an infinitely long cylinder and a spheroidal shape particle (Sharma and Roy 2000). In contrast, the HGPF yielded elevated values at near forward scattering angles for large particles, $g \sim 0.95$, $m = 1.107$ (the parameters relevant for a spherical biological scatterer) (Liu 1994).

A comparison of phase function for monodispersion of spherical particles for $g = 0.993$ with refractive indices and volume concentration typical of soft biomedical tissues was performed by Sharma and Banerjee (2003). It was shown that the HGPF reproduces the average shape of the oscillating Mie phase function reasonably well.

The oscillations occurring in the monodispersion or a single particle Mie phase function smooth out for scattering by a collection of size distributed particles. The HGPF could then be expected to provide improved agreement with the true phase function. Numerical comparisons however, lead to different conclusions in different contexts. For haze and clouds, use of the HGPF in computing angular scattering gives good results for thick layers (Hansen 1969). The study of scattering characteristics of clear sky cumulus showed that the errors in using the HGPF can be very large (Zhao and Sun 2010). It is found to be not very accurate for grain-size distributions relevant to interstellar medium (see, for example, Draine (2003) and Bianchi et al. (1996)). For oceanic waters, the HGPF underestimated the commonly used Petzold phase function (Petzold 1972) in the near-forward scattering (Mobley et al. 2002). In biomedical contexts, the HGPF shows significant discrepancies compared to solutions of Maxwell equations for a erythrocyte (Reynolds and McCormick 1980). The same has been found to be true for a biomedical tissue consisting of polydisperse scatterers (Sharma and Banerjee 2003). However, use of these phase functions in Monte Carlo simulations of tissue diffuse reflectance and fluence shows good agreement with those obtained by using the Mie scattering phase function (Sharma and Banerjee 2003). In contrast, for polydisperse aerosols its use in Monte Carlo simulations show significant underestimation of the transmittance (Bai et al. 2011). On the basis of error defined as,

$$\sqrt{\frac{\sum_{\theta=0}^{180}\left(p(\theta) - p_{HGPF}(\theta)\right)^2}{181}}, \tag{4.34}$$

it was concluded in the context of scattering by aerosols that the use of the HGPF for multiple scattering simulations provides adequate accuracy only if scattering is close to isotropy.

Significant improvement in the Monte Carlo simulations of light transport in tissues is achieved if g in the HGPF is replaced by g_a (Zolek et al. 2008):

$$g_a \equiv g + a + b\exp\left[-\frac{\ln(2)}{\ln^2(e)}\ln^2\left(\frac{(g-c)(e^2-1)}{de} + 1\right)\right], \tag{4.35}$$

with $a = -1.895,\ b = 2.072,\ c = 0.607,\ d = 3.416,\ e = 0.294$.

Interestingly, the HGPF yields an analytic expression for the probability of

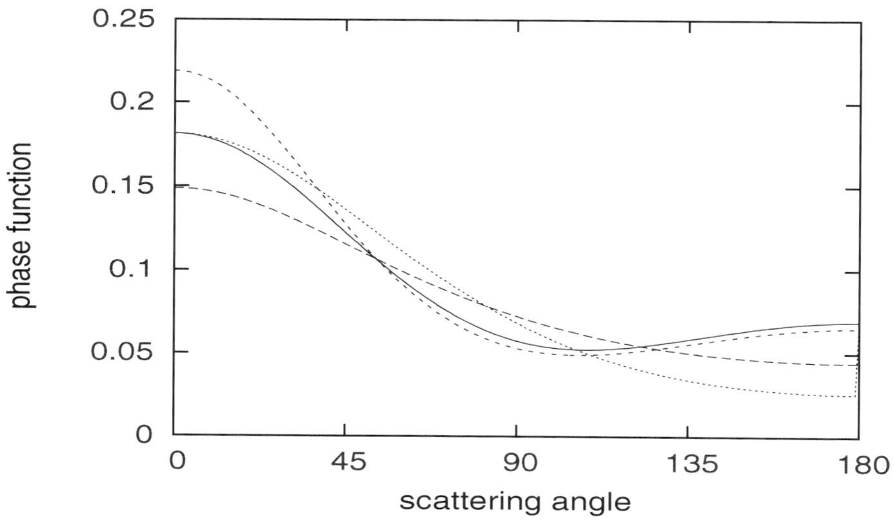

FIGURE 4.2: A comparison of some approximate small particle phase functions with the Mie phase function ($x = 1.0,\ m = 1.5$). Solid line: Mie phase function, large dashed line: HGPF, small dashed line: CASPF, dotted line: LPF. The figure was first published in Sharma et al. (1998).

backscattering:

$$B = 2\pi \int_{-1}^{0} p(\mu, g)\mu \, d\mu \;=\; \frac{1-g}{2g}\left[\frac{1-g}{\sqrt{1+g^2}} - 1\right]. \tag{4.36}$$

The measurement of B has been found to be useful for retrieving the refractive index of bulk samples (Sahu and Shanmugam 2015).

4.5.2 Improvements over the HGPF

To extend the region of applicability of the HGPF, Cornette and Shanks (1992) and Liu and Weng (2006) modified it in such a way that:

(i) It remains a one parameter phase function,
(ii) In the limit $g \to 1$ it goes to the HGPF and
(iii) In the limit $g \to 0$ it reduces to the Rayleigh phase function (RPF).

This modified HGPF reads:

$$p_{CASPF}(\mu, g) = \frac{3}{2}\frac{1+\mu^2}{2+g^2} p_{HGPF}(\theta, g) = \frac{3}{8\pi}\frac{1-g^2}{2+g^2}\frac{1+\mu^2}{(1+g^2-2g\mu)^{3/2}}. \tag{4.37}$$

The subscript $CASPF$ refers to the Cornette and Shanks phase function. It has been referred to as the merged Rayleigh–Henyey–Greenstein phase function also (Wienman and Kim 2007). For $g = 0$, (4.37) becomes

$$p_{CASPF} = 3(1+\mu^2)/16\pi, \tag{4.38}$$

which is nothing but the Rayleigh phase function (see 4.3.2). The asymmetry parameter for the CASPF is,

$$\langle\mu\rangle = g\frac{3(4+g^2)}{5(2+g^2)}, \tag{4.39}$$

where,

$$g = \frac{5}{9}\langle\mu\rangle - \left(\frac{4}{3} - \frac{25}{81}\langle\mu\rangle^2\right)y^{-1/3} + y^{1/3}, \tag{4.40}$$

and

$$y = \frac{5}{9}\langle\mu\rangle + \frac{125}{729}\langle\mu\rangle^3 + \left(\frac{64}{27} - \frac{325}{243}\langle\mu\rangle^2 + \frac{1250}{2187}\langle\mu\rangle^4\right)^{1/2}. \tag{4.41}$$

Numerical comparisons of the CASPF with the Mie phase function show it to be in good agreement only for $g \ll 1$. Similar conclusions have been drawn by Toublanc (1996).

An extensively used phase function in atmospheric applications is the Deirmendjian phase function (Deirmendjian 1969). An analytic form, which closely reproduces the appearance of tabulated Deirmendjian phase function for specific wavelengths and haze types, is the Neer–Sandri phase function (Fishburne et al. 1976):

$$p_{NSPF}(\theta) = \frac{1-g^2}{4\pi}\left[\frac{1}{(1+g^2-2g\cos\theta)^{3/2}} + f(3\mu^2-1)\right]. \tag{4.42}$$

The added term on the right-hand side adds to backscattering. It is symmetric around $\theta = 90$ deg. Therefore, the normalization as well as g remain unaltered. The parameter f is meant to adjust the relative strength of forward and backward scattering. Fishburne et al. (1976) figured f by imposing the condition that the phase function is positive for all scattering angles. Thus, equation (4.42) gives,

$$p_{NSPF}(\theta) = \frac{1-g^2}{4\pi}\left[\frac{1}{(1+g^2-2g\cos\theta)^{3/2}} + \frac{g(3\mu^2-1)}{2|\cos\theta_0|(1+g^2+2g\cos\theta_0)^{5/2}}\right], \tag{4.43}$$

where θ_0 corresponds to the minimum value of the phase function. It is found that for $\cos\theta_0 = 1/7$ the phase function remains positive everywhere and looks similar to Deirmendjian's phase function. With minor alterations, this phase function has also been adapted to aerosol scattering (Ben-Zvi 2007) and in modelling of optical wireless scattering communication channels (Liu et al. 2015). Riewe and Green (1978) have shown that by a judicious choice of f, the Mie phase function can be reproduced quite accurately.

The Gegenbauer Kernel phase function (GKPF) (Reynolds and McCormick 1980) is a two-parameter phase function. The HGPF appears as a particular case of the GKPF:

$$p_{GKPF}(\mu) = K\ (1+g^2-2g\mu)^{-(1+\epsilon)}, \tag{4.44}$$

where

$$K = \frac{1}{\pi}\frac{\epsilon g(1-g^2)^{2\epsilon}}{[(1+g)^{2\epsilon}-(1-g)^{2\epsilon}]}. \tag{4.45}$$

For $\epsilon = 1/2$, p_{GKPF} is nothing but the p_{HGPF}. In addition, like the HGPF, the GKPF also gives an analytic form for the random scattering angle (Yaroslavsky et al. 1999):

$$\cos\theta_{GKPF} = \left[\frac{1+g^2-\Lambda^{-1/2\epsilon}}{2g}\right], \tag{4.46}$$

where

$$\Lambda = \frac{2g\epsilon\xi}{K} + (1+g)^{-2\epsilon}, \tag{4.47}$$

and ξ is a random variable uniformly distributed between 0 and 1.

Numerical studies for its validity show that:

(i) Large diameter, low index particles can be reasonably approximated by the HGPF, and
(ii) Higher index particles are seen to fit better by the GKPF ($\epsilon = 3/2$ in (4.44) and (4.5)).

However, exceptions exist. For example, the angular scattering characteristics for a single red blood cell are more closely represented by the $p_{GKPF}(\theta)$ in comparison to $p_{HGPF}(\theta)$ (Reynolds and McCormick 1980, Yaroslavsky et al. 1997, Hammer et al. 1998). The same has been found to be true for the whole blood (Hammer et al. 2001). The RBCs, however, come in the category of large diameter, low index particles.

4.5.3 Sum of two phase functions

A simple way to generate the backscattering rise seen in most phase functions, is to express the phase function as a composite function of two HGPFs (Irvine 1963):

$$p_{TTHG}(\theta) = \alpha p_{HGPF}(\theta, g_1) + (1-\alpha) p_{HGPF}(\theta, -g_2), \tag{4.48}$$

$$= \sum_{n=0}^{\infty} (2n+1) \left[\alpha g_1^n + (1-\alpha)(-g_2)^n\right] P_n(\mu). \tag{4.49}$$

The subscript TTHG denotes two-term Henyey–Greenstein. A positive sign in front of g_1 results in a forward peak, while a negative sign in front of g_2 results in a backscattering rise.

For a marine environment, the following parameterization for g_2 has been obtained by Haltrin (2002):

$$g_2 = -0.30614 + 1.0006 g_1 - 0.01826 g_1^2 - 0.03644 g_1^3, \tag{4.50}$$

with

$$\langle \mu \rangle = \alpha(g_1 + g_2) - g_2, \tag{4.51}$$

and

$$\alpha = \frac{g_2(1+g_2)}{(g_1+g_2)(1+g_2-g_1)}. \tag{4.52}$$

The above parameterization is applicable in the range $0.30664 < g_1 \leq 1$.

For large biomedical particles, α was redefined by Sharma and Banerjee (2003) as:

$$\alpha = A + \frac{g_2(1+g_2)}{(g_1+g_2)(1+g_2-g_1)}. \tag{4.53}$$

The desired backscattering rise was obtained by adjusting A to $A = 0.055$. However, in a tissue model, consisting of particles of fractal size distribution, none of the phase functions, HGPF, GKPF or
TTHG, was found in good agreement with the Mie phase function. Yet, the congruity of the HGPF and the TTHG with Mie Phase functions in computation of diffuse reflectance and fluence is reasonably good.

The phase function (4.48) reproduces the experimentally observed phase function of the white matter of a neonatal brain correctly with parameter values $\alpha = 0.995$, $g_1 = 0.992$ and $g_2 = -0.93$, respectively (Kienle 2001). A special case of (4.48):

$$p_{JPF}(\theta) = \alpha p_{HGPF}(\theta, g) + \frac{1}{4\pi}(1-\alpha); \qquad 0 \leq \alpha \leq 1, \tag{4.54}$$

describes very well the scattering phase functions of human dermis (Jacques et al. 1987), aorta (Yoon 1988) and dental enamel (Fried et al. 1995).

Two hybrid phase functions that are formally similar to (4.48) are from Bevilacqua and Depeursinge (1999). The first of these is expressed as,

$$p_{BDPF1} = \alpha p_{HGPF} + \frac{3}{4\pi}(1-\alpha)\cos^2\theta; \qquad 0 < \alpha < 1. \tag{4.55}$$

The first three moments of this phase function are,

$$\langle\mu\rangle \equiv \langle\cos\theta\rangle = \alpha g, \tag{4.56}$$

$$\langle\mu^2\rangle \equiv \langle\cos^2\theta\rangle = \alpha g^2 + \frac{2}{5}(1-\alpha), \tag{4.57}$$

and

$$\langle\mu^3\rangle \equiv \langle\cos^3\theta\rangle = \alpha g^3. \tag{4.58}$$

These equations allow independent adjustment of the parameters g and α for which the phase function is positive. The other phase function by Bevilacqua and Depeursinge (1999) has been constructed by replacing p_{HGPF} in (4.55) by the binomial phase function. The resulting phase function is:

$$p_{BDPF2} = \frac{\alpha}{4\pi}\frac{N+1}{2^N}(1+\cos\theta)^N + \frac{3}{4\pi}(1-\alpha)\cos^2\theta. \tag{4.59}$$

The first two moments of this phase function are:

$$\langle\mu\rangle \equiv \langle\cos\theta\rangle = \alpha g, \tag{4.60}$$

$$\langle\mu^2\rangle \equiv \langle\cos^2\theta\rangle = \alpha g^2 + \frac{2}{5}(1-\alpha), \tag{4.60a}$$

where

$$g = \frac{N}{N+2}. \tag{4.60b}$$

Phase functions (4.55) and (4.59) cover most of the $\langle\mu\rangle$ and $\langle\mu^2\rangle$ values relevant in Mie scattering in the size parameter range 1–25. Similar phase functions have been noted in other biological tissues as well (Flock et al. 1987, Jacques et al. 1987, Marchesini et al. 1989, van der Zee et al. 1993, Chicea and Chicea 2006, Fernandez-Oliveras et al. 2012 etc.).

A modified two-term GKPF,

$$p_{MGKPF}(\theta) = \alpha p_{GKPF}(\theta) + \frac{3}{4\pi}(1-\alpha)\cos^2\theta, \tag{4.61}$$

has also been considered by Vaudelle (2017) in the context of the effective phase function.

4.5.4 Caldas–Semião approximation

A phase function for large spheres has been suggested by Caldas and Semião (2001b). The derivation employs geometrical optics and Fraunhofer diffraction, leading to a mathematically simple phase function. This phase function reads as:

$$p_{CSPFL}(\mu) = \frac{1}{4\pi}\left[a_0 + b_0\frac{2-\mu}{3-2\mu+\sqrt{8(1-\mu)}} + c_0(1+\mu)^2\exp(-d_0^2(1-\mu^2))\right], \tag{4.62}$$

where

$$a_0 = \frac{1}{2} - \frac{b}{4}(I_1+2), \qquad b_0 = \frac{2(2g-1)}{2-I_1+2I_2}, \tag{4.62a}$$

$$c_0 = \frac{Pf-a-b}{4}, \qquad d_0^2 = \frac{2c}{2g-I_2b} - \frac{1}{2}, \tag{4.62b}$$

with

$$I_2 = \frac{1}{3} - \frac{1}{4}\ln(3), \qquad I_1 = \frac{5}{2}\ln(3) - 2, \tag{4.62c}$$

and

$$Pf = \frac{1}{4}x^2\frac{Q_{ext}^2}{Q_{sca}}. \tag{4.62d}$$

The letter L in the subscript stands for large particles, g is the asymmetry parameter and $\mu = \cos\theta$. The parameters in (4.62) have been obtained by employing the conditions that:

(i) It satisfies the normalization condition (4.25),
(ii) The phase function ϕ_{CSPFL} matches with ϕ_{EX} at $\theta = 0$, and
(iii) It reproduces the asymmetry parameter g of the exact phase function correctly.

The phase function (4.62) is found to gives accurate results provided:

(i) The scatterers behave as opaque particles, and
(ii) The size parameter of the scatterers is greater than 5.

4.5.5 Biomedical specific phase functions

Most biological scatterers are large particles, scattering dominantly in the near forward direction. This prompted Liu (1994) to design phase functions constrained to reproduce the forward scattering correctly. The phase function reads:

$$p_{LPF} = K\Big(1 + \upsilon \cos\theta\Big)^{\ell}, \tag{4.63}$$

where υ and ℓ are two independent parameters. The normalization condition (4.25) gives,

$$K = \frac{\upsilon(\ell+1)}{2\pi[(1+\upsilon)^{\ell+1} - (1-\upsilon)^{\ell+1}]}. \tag{4.64}$$

The parameters υ and ℓ are the characteristic factor and the anisotropy index, respectively. A positive value of υ implies that the scattering is forward peaked. The larger the value of υ, the stronger forward peak is. For isotropic scattering $\upsilon = 0$. Negative υ indicates a backscattering rise. The anisotropy index is limited to non-negative, even integers, to ensure that the phase function remains non-negative. The larger the value of ℓ, the stronger the scattering anisotropy is.

The parameters ℓ and υ can be obtained by solving the simultaneous equations:

$$p_{LPF}(0) = K(1+\upsilon)^{\ell}, \tag{4.65}$$

and

$$g = \frac{\ell+1}{\ell+2}\left[\frac{(1+\upsilon)^{\ell+1} + (1-\upsilon)^{\ell+1}}{(1+\upsilon)^{\ell+1} - (1-\upsilon)^{\ell+1}}\right] - \frac{1}{\upsilon(\ell+2)}. \tag{4.66}$$

To determine ℓ and υ, $p_{LPF}(0)$ is taken to be the exact forward scattering phase function $p(0)$ and g is the asymmetry parameter of the to be parameterized phase function. Numerical estimates, for small particles, show that ϕ_{LPF} gives large errors away from the near-forward scattering (Figure 4.2).

For $\upsilon = 1$,

$$p_{LPF1} = \frac{1}{4\pi}\frac{\ell+1}{2^{\ell}}(1+\cos\theta)^{\ell}, \tag{4.67}$$

which may be recognized as the binomial phase function. Note that p_{LPF1} depends only on the parameter l. Nevertheless, even this simplified phase function displays better agreement with the Mie theory than the HGPF for $g = 0.9$, 0.95 and 0.99. Such forward peaked phase functions are regularly encountered in biomedical applications.

Other parameterizations for forward peaked phase functions are the delta-M phase function (Wiscombe 1977), peak truncated phase function (Hansen 1969, Potter 1970), delta-Eddington phase function (Joseph et al. 1976, Crossbie and Davidson 1985), delta-hyperbolic phase function (Haltrin 1988) and the transport phase function (Davison 1957). The last mentioned phase function has been generalized to include purely backward and low-order anisotropic scattering (Siewert and Williams 1977, Devaux et al. 1979).

Often, the scattering in biomedical tissues has been modelled as arising from scattering by refractive index fluctuations in the tissue. If $m(\mathbf{r})$ and m_0 denote the refractive index distribution of the medium and the refractive index of the background, respectively, one can define the refractive index fluctuation by,

$$\Delta m(\mathbf{r}) = (m(\mathbf{r}) - m_0)/m_0. \tag{4.68}$$

The mathematical treatment of scattering for this class of models relies on the RGA. Recall, that this approximation is also known as the Born approximation. The validity domain of this approximation from (3.70b) is,

$$\Delta m(\mathbf{r})kL \ll 1, \tag{4.69}$$

where L is the average size of the sample. When this condition is satisfied, the mean differential scattering cross section per unit volume can be written as (Ishimaru 1999, Wax and Backman 2010)),

$$\sigma(\mathbf{q}) = 2\pi k^4 \sin^2 \chi \Phi(\mathbf{q}), \tag{4.70}$$

where χ is the angle between the incident electric vector and the direction of observation $\hat{k}_s$, and

$$\Phi(\mathbf{q}) = \frac{1}{2\pi^3} \int e^{-i\mathbf{q}.\mathbf{r}} R(\mathbf{r}) d\mathbf{r}, \tag{4.71}$$

is the power spectral density of the refractive index fluctuations. A mathematical expression suitable for modelling the refractive index correlation, $R(\mathbf{r})$, in a biomedical tissue is the Whittle–Matern (WM) correlation function:

$$R(r) = \Delta m^2 \frac{2^{5/2-\nu}}{|\Gamma(\nu - 3/2)|} (r/l_c)^{\nu-3/2} K_{\nu-3/2}\left(\frac{r}{l_c}\right), \tag{4.72}$$

where Δm^2 is the variance of the refractive index, l_c is the correlation length, the parameter ν determines the shape of the function and $K_{\nu-3/2}$ is the modified Bessel function.

Substituting (4.72) into (4.71), the following expression for the mean differential scattering cross section (4.70) is obtained (Sheppard et al. 2007, Rogers et al. 2009, Wax and Backman 2010):

$$\sigma(\mathbf{q}) = \frac{2\Delta m^2 k^4 l_c^3 \Gamma(\nu)(1 - \sin^2\theta \cos^2\varphi)}{\sqrt{\pi}|\Gamma(\nu - 3/2)|} (1 + (2kl_c \sin(\theta/2))^2)^{\nu}, \tag{4.73}$$

which is related to the phase function through a simple relation,

$$p(\theta) = \sigma(\theta)/C_{sca}. \tag{4.74}$$

It is also possible to obtain analytic expressions for $\langle\mu_{sca}\rangle$, and g:

$$\langle\mu_{sca}\rangle = \frac{\Delta m^2\sqrt{\pi}(\nu-3)}{2k^2l_c^3|\Gamma(\nu-3/2)|}\Big[(1+2k^2l_c^2(2k^2l_c^2(\nu-2)-1)(\nu-3))$$

$$-(1+4k^2l_c^2)^{1-\nu}\left(1+2k^2l_c^2(\nu+1)+4k^4l_c^4(4+(\nu-3))\right)\Big], \tag{4.75}$$

and

$$g = Nr/Dr, \tag{4.76}$$

where

$$Nr = \Bigg[A^{\nu}\Bigg(3+2k^2l_c^2(\nu-4)\Big(-3-4k^2l_c^2\Big(k^2l_c^2(\nu-2)-1\Big)(\nu-3)\Big)\Bigg)$$

$$-A\Bigg(3+6k^2l_c^2(2+\nu)+8k^6l_c^6\nu\Big(10+(\nu-5)\nu\Big)+8k^4l_c^4\Big(6+(\nu-1)\nu\Big)\Bigg)\Bigg], \tag{4.77}$$

and

$$Dr = \Bigg[k^2l_c^2(\nu-4)\Bigg(A^{\nu}\Big(-1-2k^2l_c^2\Big(2k^2l_c^2(\nu-2)-1\Big)(\nu-3)\Big)$$

$$A\Big(1+2k^2l_c^2(1+\nu)+4k^4l_c^4\Big(4+(\nu-3)\nu\Big)\Big)\Bigg)\Bigg], \tag{4.78}$$

where, $A = 1+4k^2l_c^2$. For scalar wave scattering, the phase function simplifies to (Turzhitsky et al. 2010):

$$p(\theta)_{WMPF} = \frac{\hat{g}(\nu-1)\left[1-2\hat{g}\cos\theta+\hat{g}^2\right]^{-\nu}}{\pi\left[(1-\hat{g})^{2-2\nu}-(1+\hat{g})^{2-2\nu}\right]}, \tag{4.79}$$

where

$$\hat{g} = 1-\frac{\sqrt{1+4\left(kl_c\right)^2}-1}{2\left(kl_c\right)^2}, \tag{4.80}$$

and

$$kl_c = \sqrt{\hat{g}}/(1-\hat{g}), \tag{4.81}$$

and the subscript WMPF stands for Whittle–Matern phase function. If $\hat{g}$ is identified as g (anisotropy parameter), then (4.79) reduces to the HGPF for $\nu = 1.5$. For other values of ν,

$$g = \frac{(1-\hat{g})^{2-2\nu}\left(1+\hat{g}^2-2\hat{g}(\nu-1)\right)-(1+\hat{g})^{2-2\nu}\left(1+\hat{g}^2+2\hat{g}(\nu-1)\right)}{2\hat{g}(\nu-2)\left[(1+\hat{g})^{2-2\nu}-(1-\hat{g})^{2-2\nu}\right]}, \tag{4.82}$$

if $\nu \neq 2$, and

$$g = \frac{1+\hat{g}^2}{2\hat{g}} + \frac{\ln(1+\hat{g}) - \ln(1-\hat{g})}{\hat{g}\left[(1+\hat{g})^{-2} - (1-\hat{g})^{-2}\right]}, \tag{4.83}$$

for $\nu = 2$. The phase function has discontinuities at $\nu = 1$ and $g = 0$. At these values, L'Hospital's rule may be used to evaluate the phase function.

The random scattering angle, as a function of the random variable ξ, is found to be,

$$\cos\theta = \frac{1+\hat{g}^2 - \left[\xi\left((1-g)^{2-2\nu} - (1+g)^{2-2\nu}\right) + (1+g)^{2-2\nu}\right]^{\frac{1}{1-\nu}}}{2g}, \tag{4.84}$$

where ξ, as in (4.33), is a random variable uniformly distributed over the range 0 and 1.

An alternative correlation function, which is an average of exponential functions, weighted by power law distribution is,

$$\eta(l) = \eta_0 l^{3-D_f} \tag{4.85}$$

of correlation length l (Xu and Alfano 2005) is:

$$R(r) = \varepsilon^2 \int_0^{l_{max}} \exp\left(-\frac{r}{l}\right)\,\eta(l)\,dl \;=\; \varepsilon^2 \eta_0 l_{max}^{4-D_f} E_{5-D_f}\left(\frac{r}{l_{max}}\right), \tag{4.86}$$

where

$$E_n \equiv \int_0^1 e^{(-z/t)} t^{n-2}\,dt, \quad \varepsilon^2 = 4n_0^4(m-1)^2, \tag{4.87}$$

with l_{max} as the cutoff correlation length, η_0 is a constant and D_f is the fractal dimension. This correlation function leads to the following unnormalized phase function when $kl_{max} \gg 1$:

$$p(\theta) = \frac{(1+\mu^2)|S(\theta)|^2}{2k^2} \;=\; \frac{A}{\sin A}\frac{1+\mu^2}{2}\left[2(1-\mu)\right]^{-7-D_f/2}, \tag{4.88}$$

where $A = (5-d_f)\pi/2$, and

$$\mu_s' = (2\pi n_0)^{D_f-3}\frac{(11-4D_f+D_f^2)}{(D_f+1)(d_f-1)(D_f-3)}\frac{A}{2^{5-D_f}\sin A}, \tag{4.89}$$

is the corresponding reduced scattering coefficient.

4.5.6 Astrophysics specific phase functions

The HGPF, though originated in an astrophysics context, has been found to be inadequate at certain wavelengths for the size distributions relevant to the interstellar medium. A phase function that represents the interstellar dust phase function well at various wavelengths is (Draine 2003):

$$p(\theta)_{DPF} = \frac{1-g^2}{4\pi(1+g^2-2g\cos\theta)^{3/2}}\frac{1+\alpha\cos^2\theta}{1+\alpha(1+2g^2)/3}. \tag{4.90}$$

At $\alpha = 0$, $p(\theta)_{DPF}$ reduces to the HGPF and for $\alpha = 1$ it reduces to the CASPF. The parameters g and α are determined by requiring that:

(i) p_{DPF} reproduces correctly, the first moment, $\langle\cos\theta\rangle$, of the phase function and
(ii) It also reproduces the second moment $\langle\cos^2\theta\rangle$ of the phase function to be parameterized.

Implementation of these requirements, leads to the following expressions for g:

$$g = \left[(a^3+b^2)^{1/2}-b\right]^{1/3} - \left[(a^3+b^2)^{1/2}+b\right]^{1/3} + \frac{17}{9}\langle\cos\theta\rangle, \tag{4.91a}$$

if $a^3 + b^2 > 0$, and

$$g = 2|a|^{1/2}\cos\left(\frac{\Psi}{3}\right) + \frac{17}{9}\langle\cos\theta\rangle, \tag{4.91b}$$

if $a^3 + b^2 < 0$. In (4.91a) and (4.91b),

$$a = \frac{7}{3}\langle\cos^2\theta\rangle - \frac{289}{81}\langle\cos\theta\rangle^2, \tag{4.92}$$

$$b = \frac{119}{18}\langle\cos\theta\rangle\langle\cos^2\theta\rangle - \frac{4913}{729}\langle\cos\theta\rangle^3 - \frac{7}{6}\langle\cos\theta\rangle \tag{4.93}$$

and

$$\Psi = \cos^{-1}\left(\frac{-b}{|a|^{3/2}}\right). \tag{4.94}$$

The parameter α is found to be:

$$\alpha = \frac{15(\langle\cos\theta\rangle - g)}{3(3+2g^2)g - 5(1+2g^2)\langle\cos\theta\rangle}. \tag{4.95}$$

This phase function leads to an improved fit if $\alpha < 1$. For the values of $\langle\cos\theta\rangle$ and $\langle\cos^2\theta\rangle$ that lead to $\alpha > 1$, g is obtained by setting $\alpha = 1$.

The DPF constitutes significant improvement over the HGPF for interstellar dust. This can be seen in Figure 4.3, which depicts relative rms error as a function of wavelength. The relative rms error is defined as,

$$h_{rel} = \left[\frac{1}{2}\int d\mu\left(\frac{p_{app}(\mu) - p_{EX}(\mu)}{p_{EX}(\mu)}\right)\right]^{1/2}. \tag{4.96}$$

In addition to errors in the HGPF and DPF, the errors in the six-parameter phase function (SIXPPF) (see 3.3.2) have been also plotted. Note that all three phase functions give comparable errors in the ultraviolet and visible region. In the infrared region, particle size in relation to wavelength is smaller. Thus, the g values of the phase function are also smaller. For example, for $\lambda = 3.45\ \mu m$, $g = 0.00556$. In this case, it is important that the approximate phase functions agree well with the exact phase function at all scattering angles. In this respect, SIXPPF seems to give the best results. The DPF is an improvement over the HGPF in almost the entire wavelength range of interest.

The reflection from planetary surfaces is another problem of interest in astrophysics. Hapke (1981) has given a formula for planetary surface reflection. The phase function involved in this formula is a two parameter phase function. In terms of two parameters, b and c, it may be expressed as (see, for example, Hapke 1981, Mishchenko et al. 1999):

$$p(\theta_p) = \frac{1}{8\pi}\left[\frac{(1+c)(1-b^2)}{(1-2b\cos\theta_p+b^2)^{3/2}} + \frac{(1-c)(1-b^2)}{(1+2b\cos\theta_p+b^2)^{3/2}}\right] \tag{4.97}$$

where θ_p is the phase angle ($\theta_p \equiv 180 - \theta$). Equation (4.97) can be obtained from (4.48) by substituting $\alpha = (1+c)/2$ and $g_1 = g_2 = b$.

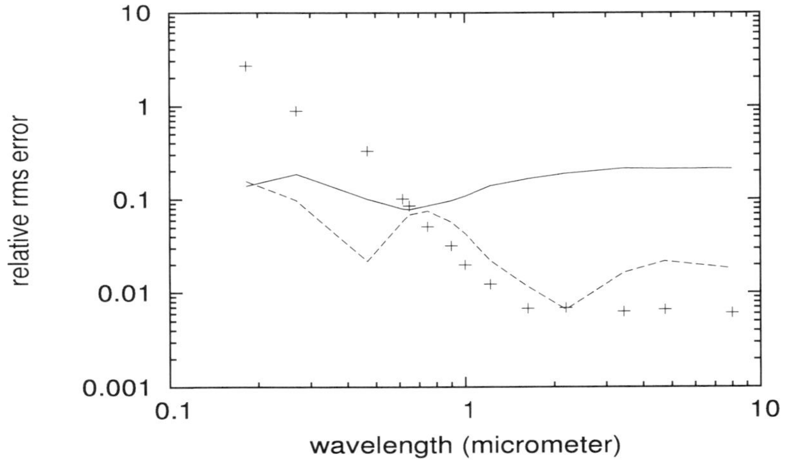

FIGURE 4.3: Relative root mean square error, as defined in (4.96), as a function of wavelength for a model of interstellar dust particles. Solid line: HGPF, dashed line: DPF and crosses: SIXPPF. The figure was first published in Sharma and Roy (2008).

The parameters b and c dependend on the material properties of the regolith. While b is restricted by the condition $0 \leq b \leq 1$, there is no restriction on c except that $p(\theta_p) \geq 0$ for all θ_p. The parameter b describes the angular width of each lobe and c describes the amplitude of the back scattered lobe relative to the forward. A positive value of c indicates greater back scattering and a negative value of c implies greater forward scattering. If b is close to 1, the lobes are high and narrow, and if $b \leq 1$, the lobes are broader and low.

It generally provides reasonably good description of the data for large irregular particles. The original Hapke model has been modified from time to time by many authors (Mishchenko 1994, Liang and Townshend 1996, Mishchenko and Macke 1997, etc.).

4.5.7 Marine environment specific phase functions

In the terminology of scattering in natural waters, the phase function is generally defined as,

$$p(\theta, \lambda) = \frac{\beta(\theta, \lambda)}{b(\lambda)}, \tag{4.98}$$

where,

$$\beta(\theta, \lambda) = \int_0^\infty \bar{\sigma}_{sca}(\theta, \lambda, a)\ f(a)da, \tag{4.99}$$

is the volume scattering function, with $\bar{\sigma}_{sca}$ as the scattering cross section averaged over shape effects, and

$$b(\lambda) = \int_0^{2\pi} \int_0^{\pi} \beta(\theta, \lambda) \sin\theta\ d\theta d\varphi, \tag{4.100}$$

is the total scattering coefficient per unit volume.

Phase functions suitable for marine environments, developed till 2007, can be found in Jonasz and Fournier (2007). Among the later developments, Sahu and Shanmugam (2015) have proposed a phase function modelled in terms of slope of the size distribution and refractive index of the particles. The range of applicable relative refractive index values are $m = 1.04$ to 1.30. The angular range 0.1-90 deg has been divided in two parts: (i) 0.1-5 deg and (ii) 5-90 deg. The phase functions have been obtained purely from numerically computed data.

The phase functions have been expressed in the angular range (i) as:

$$\log p(\theta) = P_1 (\ln(\theta))^2 + P_2 \ln(\theta) + P_3, \tag{4.101}$$

and in the angular range (ii) as:

$$\log (p(\theta)) = P_1 (\ln(\theta))^3 + P_2 (\ln(\theta))^2 + P_3 \ln(\theta) + P_4, \tag{4.102}$$

where,

$$P_l = a_l \exp(-x) + b_l(x) + c_l, \tag{4.103}$$

and

$$a_l = \frac{d_l}{y^2} + e_l \sin(5y) + f_l, \tag{4.103a}$$

$$b_l = \frac{h_l}{y^2} + i_l \sin(5y) + f_l, \tag{4.103b}$$

$$c_l = \frac{k_l}{y^2} + l_l \sin(5y) + o_l, \tag{4.103c}$$

with $x = \xi - 3$, $y = (m - 1)$ and $l = 1, 2, 3$ for (4.101) and $l = 1, 2, 3, 4$ for (4.102). The slope ξ for the size distribution $f(D)$ is defined by the relation,

$$f(D) = KD^{-\xi}, \tag{4.104}$$

with

$$K = \frac{(\xi - 1)}{D_{min}^{1-\xi} - D_{max}^{1-\xi}}, \tag{4.104a}$$

as the result of the normalization condition (4.25). D_{min} and D_{max} are the minimum and the maximum diameters, respectively, of particles in the distribution. A simulation was carried out by choosing the wavelength $\lambda = 530nm$ and the relative refractive index variation is in the range of oceanic environment (1.04–1.3). The values of the coefficients d_l, h_l k_l, e_l, i_l, l_l are different for the two cases and can be found in Sahu and Shanmugam (2015). The results from this phase function were verified by comparing them with the Petzold average particle phase function and with the measured phase function data from Sokolov et al. (2010).

The relationship between the backscattering probability B (as defined in 4.36) and m is found to be:

$$B = P_1 \ (\alpha)^{P_2}, \tag{4.105}$$

where

$$\alpha = 1/(m-1)^2, \tag{4.105a}$$

and

$$P_l = a_l(\xi - 3)^2 + b_l(\xi - 3) + c_l. \tag{4.105b}$$

The values of the constants a_l, b_l and c_l are given in Sahu and Shanmugam (2015). These are different from those defined for 0-90 deg scattering. This parameterization is valid for all values of the slope between 3 and 5. Equation (4.105) can be inverted,

$$m = 1 + \left(\frac{P_1}{B}\right)^{1/2P_2}, \tag{4.105c}$$

to give a bulk refractive index using measurable parameter ξ.

One of the most frequently used phase functions for oceanic particle scattering is the Fournier and Forand phase function (Fournier and Forand 1994, Forand and Fournier 1999). This phase function will be discussed in Section 4.6.2. Empirical, in-water, near forward scattering phase functions for randomly shaped terrigenous sediment grains can be found in Agrawal and Mikkelsen (2009).

4.5.8 Single scattering properties of snow

Analytic expressions for the single scattering properties of snow have been given recently by (Räisänen et al. 2015). Snow has been modeled as an optimized habit combination (OHC) consisting of severely rough droxtals, severely rough plane aggregates and severely distorted Koch fractals. Defining a size parameter,

$$x = x_{vp} = 2\pi\frac{r_{vp}}{\lambda}, \qquad (4.106)$$

where $r_{vp} = 0.75V/P$ is the volume to projected area equivalent radius. The single scattering properties of snow in the wavelength domain $0.199 \leq \lambda \leq 2.7\ \mu m$ and the size range $10 \leq r_{vp} \leq 2000\ \mu m$, were parameterized as follows:

The single scattering co-albedo:

$$\varpi_0 = 0.470\left[1 - \exp\left[-2.69x_{abs}\left(1 - 0.31\min(x_{abs}, 2)^{0.67}\right)\right]\right], \qquad (4.107)$$

where, the size parameter for absorption is defined as,

$$x_{abs} = \frac{2\pi r_{vp}}{\lambda} n_i n_r^2. \qquad (4.107a)$$

The relative errors, defined as $\Delta\varpi_0/\varpi_0$, are generally less than 1%. Errors larger than 10% occur only for small snow grains at $\lambda > 1.2\ \mu m$.

Asymmetry parameter:

$$g = 1 - 1.146[n_r - 1]^{0.8}[0.52 - \varpi_0]^{1.05}\left[1 + 8x_{vp}^{-1.5}\right]. \qquad (4.108)$$

It decreases with increasing n_r, increases with increasing absorption (or ϖ_0) and increasing x_{vp}. The difference from the reference value is mostly below 0.001 for $\lambda < 1.4\ \mu m$. Differences up to 0.007 were noted at strongly absorbing wavelengths. The ϖ_0 used in (4.108) is from (4.107).

The phase function:

$$p(\theta) = w_{diff}\ p_{diff}(\theta) + w_{ray}\ p_{ray}(\theta) + P_{res}(\theta). \qquad (4.109)$$

The first term on the right-hand side of (4.109) is the diffraction contribution, the second term is the ray tracing contribution, and the third term accounts for the contribution of corrections. The weight factors are,

$$w_{diff} = \frac{1}{\omega_0 Q_{ext}}, \tag{4.109a}$$

and

$$w_{ray} = \frac{\omega_0 Q_{ext} - 1}{\omega_0 Q_{ext}}. \tag{4.109b}$$

The diffraction part of the phase function is taken to be:

$$p_{diff}(\theta) = p_{HGPF}(\theta, g_{diff}), \tag{4.110}$$

where

$$g \equiv g_{diff} = 1 - \frac{0.60}{x_{vp}} = 1 - \frac{0.92}{x_p}, \tag{4.110a}$$

and $x_p = 1.535 x_{vp}$ is specific to OHC. The phase function for the ray tracing part is

$$p_{ray}(\theta) = w_1\, p_{HGPF}(\theta, g_1) + (1 - w_1), \tag{4.111}$$

where

$$w_1 = 1 - 1.53\ \max(0.77 - g_{ray}), \tag{4.112}$$

and

$$g_1 = g_{ray}/w_1, \tag{4.113}$$

with

$$g_{ray} = \frac{g - w_{diff}\ g_{diff}}{w_{ray}}, \tag{4.114}$$

as the asymmetry parameter for the ray tracing part. The sum of the first two terms provides a reasonably good approximation. However, further improvement can be obtained, if required, by considering the residual term.

Simpler expressions for the phase function of ice clouds have been obtained by Kokhanovsky (2008a), who also fit the experimental measurements of scattered light intensity for ice clouds.

4.6 Some distribution specific analytic phase functions

4.6.1 Rayleigh phase function for modified gamma distribution

An approximate analytic phase function for an aerosol population of size distribution

$$f(r) = Ar^a e^{-br}; \quad a \le 0, \quad b > 0, \tag{4.115}$$

has been derived by Kocifaj (2011). In (4.115), r is the equivalent radius of the aerosol particles. By varying the free parameters a and b, the modal radius ($r_m = a/b$) as well as the half width σ of the size distribution can be adjusted. The coefficient A is proportional to the volume concentration of the particles.

Scattering by a size distributed population can be obtained by integrating the product of

$$\frac{I_\lambda(\theta, r)}{I_{0,\lambda}} = \left| \frac{3}{4\pi} \left(\frac{m^2 - 1}{m^2 + 1} \right) \right|^2 \times \frac{1 + \cos^2\theta}{(1 - \cos\theta)^2} \left(\frac{\sin q - q\cos q}{q} \right)^2, \tag{4.116}$$

and $f(r)$ over all values of the particle radii. Equation (4.116) gives scattering by a particle in the RGA, where $I_{0,\lambda}$ is the incident intensity and $q = (4\pi r/\lambda)\sin(\theta/2)$ is the momentum transfer. It is expected to be applicable only to small and intermediate size particles. However, when integrated over a large ensemble of randomly sized particles, the resulting phase function is very similar to the realistic phase function.

The integration yields the following analytic form for the relative intensity of the scattered radiation:

$$I_\lambda(\theta, a, b) \propto \frac{1 + \cos^2\theta}{(1 - \cos\theta)^2} \Bigg[L_\lambda((a+1)(a+2), b, \theta)) + \left(\frac{L_\lambda(0, b, \theta)}{L_\lambda(4, b, \theta} \right)^{\frac{a+3}{2}} R_\lambda(a, b, \theta) \Bigg], \tag{4.117}$$

where

$$R_\lambda(a, b, \theta) = \sum_{i=1}^{3} 2^{f_i} \Big[f_i \sin G_{\lambda,i}(a, b, \theta) + f_{i+1} \cos G_{\lambda,i}(a, b, \theta) \Big] L_i^{\frac{3-i}{2}}(4, b, \theta) \times \prod_{j=0}^{i-1} (-1)^{i+j} \frac{a+j}{a}, \tag{4.118}$$

with

$$G_{\lambda,i}(a, b, \theta) = (a+i)\tan^{-1}\left(\frac{2}{\sqrt{L_\lambda(0, b, \theta)}} \right), \tag{4.119}$$

and

$$L_\lambda(\xi, b, \theta) = \xi + \left(\frac{\lambda b}{4\pi \sin(\theta/2)}\right)^2, \quad f_i = \frac{(-1)^i + 1}{2}. \tag{4.120}$$

Numerical tests show that introducing a modified,

$$L_\lambda^{mod}(\xi, b, \theta) = \left(\frac{2|m| - 1}{m} \frac{\lambda b}{4\pi \sin(\theta/2)}\right)^2 + \xi, \tag{4.121}$$

usually leads to better accuracy.

It may be noted that as $\theta \to 0$, the factor $(1 - \cos\theta)$ in the denominator of (4.117), approaches zero. Nonetheless, it can be shown that the limiting value of $I_\lambda(a, b, \theta)$ as $\theta \to 0$ is:

$$\lim_{\theta\to 0} I_\lambda(a, b, \theta) = \frac{8}{L_\lambda^2(0, b, \pi)} \left[\frac{a + 1)(a + 2)}{3} \left(\frac{(a+3)^4}{8} - \frac{(a+2)^4}{10} + (a+2)^2\right) + \frac{(a+1)^4}{6}\left(\frac{(a+1)^2}{30} - 1\right)\right]. \tag{4.122}$$

The phase function can now be obtained if C_{sca} is known. However, no simple algebraic solution exists for C_{sca}. Nevertheless, by numerical experimentation a simple analytic formula obtained is,

$$C(a, b, \lambda)_{sca} = \frac{\pi}{2}(1 + \gamma a)^4 \left(\frac{8\pi}{\lambda b}\right)^{4\gamma\left(1 - \frac{a}{180}\right)}, \tag{4.123}$$

which is accurate to within 5% in the intervals $1 \leq a \leq 15$, b, $5 \leq b \leq 40\ \mu m^{-1}$, and $0.4 \leq \lambda \leq 0.8\ \mu m$. The scattering phase function can then be written as,

$$p(a, b, \theta) = \frac{1}{2\pi} \frac{I(a, b, \theta)}{C_{sca}(a, b, \lambda)}. \tag{4.123a}$$

The results from this phase function have been contrasted with those from the Mie theory and the HGPF. Nonabsorbing particles with refractive index 1.61 were considered. The size distribution parameters were $a \approx 0$ and $b = 20\ \mu m^{-1}$ and $\lambda = 400\ nm$. It was found that while this scattering phase function simulates the forward scattering quite well; it underestimates both side and backward scatterings.

Unfortunately, g also cannot be derived in a simple analytic form. Nevertheless, an approximate analytic form has been obtained (Kocifaj 2011):

$$g \cong \frac{\cos^2 G(a, \lambda b)}{1 + G^2(a, \lambda b)}, \tag{4.124}$$

where

$$G(a, \lambda b) = \frac{10\gamma^2 + \lambda b}{8\pi} \frac{8}{10 + 5a/(\gamma^2\pi^2)}, \tag{4.125}$$

and $\gamma = 0.577$ is Euler's constant. Equation (4.124) is accurate to within 1% for the most typical cases of a (ranging from 0 to 15) and b (ranging from 5 to 40 μm^{-1}. The wavelength interval is 0.4 to 0.8 μm^{-1}.

4.6.2 Junge size distribution

Fournier and Forand (1994) derived an approximate analytic phase function in the framework of the ADA for an ensemble of particles that have a Junge type (hyperbolic) size distribution.

A Junge cumulative size distribution is described by $N(r)$, the number of particles per unit volume with size greater than r (volume-equivalent spherical radius), and is proportional to $r^{-\alpha}$. The number α determines the slope of the size distribution when $\log N(r)$ is plotted vs. $\log(r)$. Oceanic particle size distributions typically have α values between 3 and 5 and refractive index in the range $1.0 \leq m \leq 1.35$.

In its latest form this phase function is given by (Jonasz and Fournier 2007),

$$p_{FFPF}(\theta) = \frac{1}{4\pi(1-\delta)^2\delta^\nu}\left[\nu(1-\delta) - (1-\delta^\nu) + \left(\delta(1-\delta^\nu) - \nu(1-\delta)\right)\sin^{-2}\left(\frac{\theta}{2}\right)\right] + \frac{1-\delta_{180}^\nu}{4(\delta_{180}-1)\delta_{180}^\nu}(3\cos^2\theta - 1), \tag{4.126}$$

where

$$\nu = \frac{3-\alpha}{2}, \tag{4.127}$$

and

$$\delta = \frac{4}{3(n_r-1)^2}\sin^2\left(\frac{\theta}{2}\right), \tag{4.128}$$

with δ_{180} as the value of δ at $\theta = 180\,\text{deg}$. The subscript $FFPF$ is the abbreviation for the Fournier and Forand phase function. Equation (4.126) can be integrated to yield,

$$B = 1 - \frac{1 - \delta_{90}^{\nu+1} - 0.5(1-\delta_{90}^\nu)}{(1-\delta_{90})\delta_{90}^\nu}. \tag{4.129}$$

as the backscattering function.

The errors in commonly used phase functions for single component polydisperse seawater systems have been assessed by You-Wei et al. (2012). It was found that while both the HGPF as well as the two-term HGPF (TTHG) agree well with the theoretical ones for small particles, the FFPF can be used in the case of suspensions with large suspended particles. The fitting accuracy of the HGPF was found to be the worst. A comparison of numerical and analytical radiative-transfer solutions for the plane albedo using this phase function and the Rayleigh phase function (RPF) can be found in Sokoletsky et al. (2009).

4.7 Extinction by randomly oriented monodisperse particles

4.7.1 Cylinders

Scattering by randomly oriented finite circular cylinders has been treated by many authors within the framework of the ADA (Aas 1984, Liu et al.1998, Sun and Fu 2001). Liu et al. (1998) have contrasted the ADA predictions against the T-matrix computations of absorption and scattering efficiencies. The concurrence between the ADA and the T-matrix for ice cylinders was found to be very good. If L and a denote the length and radius, respectively, it was demonstrated that the efficiency factors were not sensitive to the ratio $2a/L$ if this ratio is less than 0.1. This implies that a cylinder can be treated as infinitely long if $a/2l < 0.1$.

The following simple expressions for efficiency factors of a thin cylinder have been obtained by making suitable expansions in the powers of ρ and κ (Aas 1984),

$$Q_{ext} \simeq 2\kappa + \frac{4}{3}(\rho^2 - \kappa^2) - \frac{3}{2}(\rho^2\kappa - \frac{\kappa^3}{3})\ln(L/2a), \tag{4.130}$$

$$Q_{abs} \simeq 2\kappa - \frac{8}{3}\kappa^2 + 2\kappa^3 \ln(L/2a), \tag{4.131}$$

and

$$Q_{sca} \simeq \frac{\pi^2}{8}(\rho^2 + \kappa^2) - \frac{3}{2}(\rho^2\kappa + \kappa^3)\ln(L/2a). \tag{4.132}$$

A parameterization for randomly oriented aggregates of finite ice cylinders at millimeter wavelengths has been given by Wienman and Kim (2007):

$$\frac{Q(\rho)}{\rho} = \frac{0.40\rho^{2.1}}{(1 + 0.031\rho^{4.1})}, \tag{4.133}$$

$$g = \frac{0.30\rho^{2.1}}{(1 + 0.18\rho^{2.6})}, \tag{4.134}$$

and

$$p(180) = \frac{3(1-g)}{(2+g^2)(1+g)^2}. \tag{4.135}$$

In the above equations α is the aspect ratio $(L/2a)$ and the phase delay is $\rho = 2\pi\Delta(m-1)\nu/c$ with $\Delta = 4\pi L/\pi(1+\alpha/2)$. The above parameterizations are valid for the range $0.1 \leq \alpha \leq 0.6$. The refractive index was taken to be $n_r = 1.78$ and $n_i = 0.004$.

For infinitely long cylinders, the averaging over random orientations of efficiencies factors in the ADA has been done by Fournier and Evans (1996):

$$\bar{Q}^{ext}_{ADA} \sim 2 + e^{-2\kappa x}\left[\frac{4}{3}\rho^2\left(1 - \frac{\rho\pi}{4}\left[J_0^2(\rho/2) - \frac{2}{\rho}J_0(\rho/2)J_1(\rho/2)\right.\right.\right.$$
$$\left.\left.\left. + \left(1-\frac{2}{\rho}\right)J_1(\rho/2)\right]\right) - 2\right]T. \tag{4.136}$$

where $\rho = \rho_{ADA} = 2x(n_r - 1)$, and

$$T = 2 - \exp(\bar{Q}_{edge}/2). \tag{4.137}$$

The contribution of the edge effects can be obtained by using a formula given by Jones (1957). For randomly oriented infinite cylinders, it yields,

$$\bar{Q}_{edge} = \frac{1.15959c_0}{x^{2/3}}. \tag{4.138}$$

In general, the function c_0 is very complex. For soft particles, however, it can be approximated as $c_0 = 0.99613$. A semi-empirical modification to extend the validity of (4.138) to smaller ρ values has been given by Fournier and Evans (1996):

$$\bar{Q}_{edge} = \frac{1.15959c_0}{x^{2/3} + \frac{\pi}{4|m-1|}}. \tag{4.139}$$

Errors are less than 5% for medium- and large-sized particles if $n_r \geq 1$ and $0 \leq n_i \leq 3$.

Extension of the above expressions to the scattering by elliptic cylinders is straightforward (Fournier and Evans 1996).

4.7.2 Spheroids and ellipsoids

Angular averaging of extinction efficiency for a spheroid leads to the following expression in the ADA (Evans and Fournier 1994):

$$\bar{Q}^{ADA}_{ext} = 2 + 4(I_1 - I_2)/j(0), \tag{4.140}$$

where,

$$I_1 = A\left[e^{-c}\left[\left(1+\frac{1}{C}\right)\frac{F_2(C)}{C} - \left(1+\frac{2}{C}\right)\frac{F_1(C)}{C^2}\right] + \left(1+\frac{2}{C}\right)\frac{F_1(0)}{C^2} + \frac{F_2(0)}{C^2}\right], \tag{4.141}$$

$$I_2 = \frac{j(0)}{C^2} + \frac{A}{C^2}\left[\frac{1}{\omega(0)} - \frac{1}{\omega(\pi/2)}\right] + \frac{2A}{C^3}\ln\left[\frac{B\omega(\pi/2)}{[B+j(0)]\omega(0)}\right], \quad (4.142)$$

with

$$A = \gamma B[B + j(0)], \quad (4.143a)$$

$$B = \frac{\omega(0) - \omega(\pi/4)}{\gamma + [\omega(\pi/4) - \omega(\pi/2)]/j(\pi/4)}, \quad (4.143b)$$

$$C = \omega(0) - \gamma B, \qquad \gamma = \frac{\omega(\pi/2) - \omega(0)}{j(0)}, \quad (4.143c)$$

$$\omega = ib\left[\frac{2r}{p}\left(\frac{p^2\cos(\vartheta) + s\sin(\vartheta)}{p^2\cos^2(\vartheta) + q^2\sin^2(\vartheta) + 2s\cos(\vartheta)\sin(\vartheta)}\right)\right] \times [m - \cos(\vartheta)], \quad (4.143d)$$

$$\cos(\vartheta) = \frac{s^2 + p^2\Delta}{m(p^4 + s^2)}, \qquad \sin(\vartheta) = \frac{s(\Delta - p^2)}{m(p^4 + s^2)}, \quad (4.143e)$$

$$\Delta = [m^2(p^4 + s^2) - s^2]^{1/2}, \quad s = (p^2q^2 - r^2)^{1/2},$$

$$q = [r^2\cos^2\theta + \sin^2\theta]^{1/2}, \quad p = \sqrt{\cos^2\theta + r^2\sin^2\theta}, \quad r = a/b, \quad (4.143f)$$

$$j(\theta) = \frac{1}{2}\left[\cos\theta\left(1 - g^2\cos^2\theta\right)^{1/2} + \frac{\sin^{-1}[g\cos\theta]}{g}\right], \quad (4.143g)$$

for prolate spheroids and

$$j(\theta) = \frac{r}{2}\left(\cos\theta\left[1 + f^2\cos^2\theta\right]^{1/2} + \frac{\ln(f\cos\theta + [1 + f^2\cos^2\theta]^{1/2}}{f}\right), \quad (4.144)$$

for oblate spheroids,

$$F_n(C) = \frac{E_n[\omega(0) - C]}{(\omega(0) - C)^{n-1}} - \frac{E_n[\omega(\pi/2) - C]}{[\omega(\pi/2) - C]^{n-1}}, \quad (4.145)$$

where E_n is the n-th order exponential integral and $r = a/b$ is the aspect ratio. α is the length of the semi-axis of rotation ($a = 2\pi\alpha/\lambda$) and β is the length of the other axis of the spheroid ($b = 2\pi\beta/\lambda$) and θ is the angle between the incident radiation and the α or a axis. For large spheroids, the edge effect can be expressed as an analytic multiplicative factor:

$$\bar{Q}_{large} = \bar{Q}_{ADA}\bar{T}. \quad (4.146)$$

Equation (4.146) has been found to gives good results for all size parameters, aspect ratios from 0.2-5, $n_r > 1$, $n_i > 0$. If high precision is not required, these formula are much more economical than the T-matrix method for obtaining the extinction efficiency factor.

As was noted in Section (3.7.8), the integral light scattering characteristics of an ellipsoid are equivalent to the integral characteristics of a sphere of radius h given by (3.189). The averaging procedure over all orientations can be achieved, to a good degree of accuracy, if one uses (Paramonov 1994),

$$< h >= \frac{3\mathcal{V}}{2 < \Sigma >}. \tag{4.147}$$

Here $< \Sigma >$ is the average value of the geometrical cross section. Further, since $< \Sigma >= S/2$ for randomly oriented convex particles, one can write $< h >= 3V/S$ for randomly oriented ellipsoidal particles. Here, S is the surface area of the particles. The formula $a =< h > /2$ has been used for many types of nonspherical particles.

4.7.3 Arbitrary shapes

Bryant and Latimer (1969) have shown that the scattering by a randomly oriented particle of volume V and a projected area P is equivalent to the scattering by a cylinder of identical volume with a diameter V/P. The incident radiation is normal to the base of the cylinder. This approximation has been referred to as the simplified ADT or the SADT. The extinction and absorption efficiency factors of this particle can, therefore, be expressed as (Sun and Fu 2001):

$$Q_{SADA}^{ext} = \left[2 - 2\exp(-k\frac{V}{P}n_i)\cos[k\frac{V}{P}(n_r - 1)]\right], \tag{4.148a}$$

and

$$Q_{SADA}^{abs} = \left[1 - \exp(-2k\frac{V}{P}n_i)\right]. \tag{4.148b}$$

For a randomly oriented circular cylinder,

$$\frac{V}{P} = \frac{2aL}{(a + L)}, \tag{4.149}$$

and for a randomly oriented hexagonal column,

$$\frac{V}{P} = \frac{2\sqrt{3}aL}{a\sqrt{3} + 2L}. \tag{4.150}$$

In either case a is the base radius and L is the height of the cylinder. Although SADT is a simple way of computing optical efficiencies for randomly oriented particles, it can also lead to large errors and hence needs to be applied selectively (Liu et al. 1996, Fu et al. 1999, Sun and Fu 2001).

More recently, a statistical interpretation of the ADA has been used to tackle particles of irregular shape. It can be seen from (3.160) and (3.161) that the numerical outcome from the two equations should be independent of the order in which the contributions to the integrals are calculated. Therefore, it is possible to cast the integration in terms of a probability function $f(l)dl$ in the following form:

$$Q_{ext}^{ADA} = 2\text{Re}\left[\int_0^{l_{max}} \left[1 - e^{ikl(m-1)}\right] f(l)dl\right], \tag{4.151}$$

and

$$Q_{abs}^{ADA} = \int_0^{l_{max}} \left[1 - e^{-kln_i}\right] f(l)dl, \tag{4.152}$$

where $f(l)$ is the chord length distribution function and $f(l)dl$ gives the probability of chord length between l and $l + dl$. The normalization condition is

$$\int_0^{l_{max}} f(l)dl = 1. \tag{4.153}$$

If $l_{min} \neq 0$, the lower limit of integration may be taken as l_{min}. It can be verified that the chord length distribution for a sphere is $f(l) = l/2$ with l between 0 and 2.

It may be noted that the percentage of particle area corresponding to a specific geometric path does not change with size. It depends only on the morphology of the particle. If $f_0(l)$ denotes the chord distribution for one particle of unit size, the ray distribution of a particle of size L can be obtained by scaling old l by l/L. Hence,

$$f(l)dl = \frac{1}{L} f_0(l/L)dl. \tag{4.154}$$

Some CLD functions and their moments for simple shapes have been examined by (Mäder 1980), Coleman (1981), Kellerer (1984), Gille (2000), Xu (2003), Xu et al. (2003), Yang et al. (2004), Jacquier and Gruy (2008a, 2008b), etc. A review on the topic can be found in Xu and Katz (2008). Analytic expressions have been obtained for spheres, spheroids, randomly oriented spheroids, cylinders, randomly oriented cylinders, double spheres and aggregates of spheres.

For an isolated ellipsoid, it has been shown that (Xu 2003, Xu et al. 2003):

$$f_0(l) = \frac{1}{2r^2b^2}(\sin^2\Psi + r^2\cos^2\Psi)lH\left[\frac{2rb}{(\sin^2\Psi + r^2\cos^2\Psi)^{1/2}} - l\right] \quad l \leq 0, \tag{4.155}$$

where $H(x)$ is the Heaviside step function and Ψ is the orientation of the spheroid. Further, if the size distribution is given as

$$f(a) = \frac{1}{\sqrt{2\pi}\sigma a}\exp\left[\frac{-\ln^2(a/a_m)}{2\sigma^2}\right], \tag{4.156}$$

the chord length distribution is given as

$$f(l) = \frac{\left(\frac{\sin^2\Psi}{r^2} + \cos^2\Psi\right) l \ \times erfc\left[\left(\frac{1}{\sqrt{2\sigma}}\right)\ln\left[\left(\frac{\sin^2\Psi}{r^2} + \cos^2\Psi\right)l/(2a_m)\right]\right]}{4a_m^2\exp(2\sigma^2)}, \tag{4.157}$$

where $erfc(x)$ is the complementary error function.

For an ensemble of particles, (4.151) can be expressed as

$$Q_{ext}^{ADA} = 2\text{Re}\left[1 - F(k(m-1))\right], \tag{4.158}$$

where

$$F(p) = \int e^{ipl} f(l)dl, \tag{4.158a}$$

is the Fourier transform of the chord length distribution (CLD) function (Malinka 2015). In the same way the absorption efficiency factor becomes,

$$Q_{abs}^{ADA} = 1 - L(p), \tag{4.159}$$

where

$$L(p) = \int e^{-pl} f(l)dl, \tag{4.159a}$$

is the Laplace transform of the CLD function $f(l)$.

Both the random orientation of particles and the size distribution, both are expected to wash out the characteristic features of an individual particle. The probability density function is then nearly Gaussian in nature. Thus, for randomly oriented polydisperse particles, it is possible to approximate,

$$f(l) = \frac{1}{\sqrt{2\pi\sigma}}\exp\left[-\frac{(l-\mu)^2}{2\sigma^2}\right]. \tag{4.160}$$

This with (4.150) and (4.151) yields,

$$Q_{ext}^{GRA} = 2 - 2\cos\left[k(n_r-1)(\mu - k\sigma^2 n_i)\right]\exp\left[-k\mu n_i - \frac{k^2\sigma^2[(n_r-1)^2 - n_i^2]}{2}\right]. \tag{4.161}$$

and

$$Q_{abs}^{GRA} = 1 - \exp\left[-2kn_r(\mu - kn_i\sigma^2)\right]. \tag{4.162}$$

The approximation has been termed the Gaussian ray approximation (GRA).

If the particle size and refractive index is such that, $kl(m-1) \ll 1$ and $kln_i \ll 1$, then (4.161) and (4.162) reduce to following equations,

$$Q_{ext}^{GRA} = 2kn_i\langle l\rangle + k^2[(n_r-1)^2 - n_i^2]\langle l^2\rangle, \tag{4.163}$$

and

$$Q_{abs}^{GRA} = 2kn_i\langle l\rangle - 2k^2n_i^2\langle l^2\rangle, \tag{4.164}$$

where $\langle l\rangle = \mu$ and $\langle l^2\rangle = \mu^2 + \sigma^2$. Equations (4.163) and (4.164) are in agreement with corresponding expressions in the RGA.

Numerical comparisons of absorption efficiency for spheres of $m = 1.05 + i0.0005$ calculated using GRA differ from Mie results at most by 2%. The extinction efficiency factors in the GRA, however, agree with the exact results only for intermediate size spheres. The maximum deviation compared with the ADA is 3.5%, whereas this is 7% for the GRA.

The scattering cross section of aggregates of nonabsorbing primary particles has been examined by Jacquier and Gruy (2008a, 2008b). Denoting the N particle aggregate by subscript N, it is straightforward to express the orientation averaged cross section C_N^{ADA}, as

$$\langle C_N^{ADA}\rangle = 2a^2\ \langle S_p\rangle \int_0^{l_{max}} [1 - \cos xl(m-1)]\ f(l)dl. \tag{4.165}$$

The $\langle S_p\rangle$ is the projected area averaged over all orientations. For aggregates, knowing this chord length distribution is not straightforward. Nevertheless, an algorithm for calculation of chord length distribution has been provided. This algorithm, while suitable for ordered aggregates, is not very good for disordered aggregates (Gruy 2014).

An alternate form for the extinction and absorption efficiency factors has been suggested by Yang et al. (2004). Defining $\tilde{l} = l/l_{max}$, the extinction and the absorption efficiency factors can be cast as,

$$Q_{ext}^{ADA} = 2\mathrm{Re}\left[\int_0^1 \left[1 - e^{ikl_{max}\tilde{l}(m-1}\right] f(\tilde{l})d\tilde{l}\right], \tag{4.166}$$

and

$$Q_{abs}^{ADA} = \int_0^1 \left[1 - e^{-kl_{max}\tilde{l}n_i}\right] f(\tilde{l})d\tilde{l}. \tag{4.166a}$$

Further, a cumulative distribution of $\bar{l}$ may be defined as:

$$q(\bar{l}) = \int_0^{\bar{l}} f(\bar{l})d\bar{l}. \tag{4.167}$$

It is then possible to write the expressions (4.166) and (4.166a) as,

$$Q_{ext}^{ADA} = 2\mathrm{Re}\left[\int_0^1 \left[1 - e^{ikl_{max}\bar{l}_q(m-1)}\right] dq\right], \tag{4.168}$$

and

$$Q_{abs}^{ADA} = \int_0^1 \left[1 - e^{-kl_{max}\bar{l}_q n_i}\right] dq, \tag{4.168a}$$

where $\bar{l}_q = \bar{l}(q)$. That is the value $\bar{l}$ corresponding to a given q. For a homogeneous sphere $\bar{l}_q = q^{1/2}$. Equations (4.168) and (4.168a), then comply with the standard ADA formula.

4.8 Extinction and scattering efficiencies by a polydispersion of spheres

4.8.1 Modified gamma size distribution in the ADA

A size distribution, representing a wide range of practical situations, is the modified Gamma distribution:

$$f(a)da = \frac{N_t}{\Gamma(s)}\left(\frac{a}{a_s}\right)^{s-1} e^{-a/a_s} d(a/a_s), \tag{4.169}$$

where a_s is the "characteristic radius" and s is the measure of variance of the distribution. The characteristic radius a_s is related to the total area S of the particles via the relation,

$$S = \int_0^\infty \pi a^2 f(a)da = N_t \pi a_s^2 s(s+1). \tag{4.170}$$

Using the relation,

$$\int_0^\infty e^{-za} a^p f(a)da = N_t a_s^p \frac{1}{(za_s+1)^{s+p}} \frac{\Gamma(s+p)}{\Gamma(s)}, \tag{4.172}$$

analytic expressions for extinction and absorption coefficients can be obtained in a straightforward way (Flatau 1992).

Casperson (1977) has given an analytic expression for the extinction coefficient for a more general size distribution in the form:

$$\eta(a) = C_n a^s \exp\left[-(a/a_s)^r\right], \tag{4.173}$$

where $a_s = a_{max}(r/s)^{1/r}$, and the normalization constant is,

$$C_n^{-1} = (a_s)^{s+1} \frac{\Gamma[(s+1)/r]}{r}. \tag{4.173a}$$

For a nonabsorbing particle, the extinction coefficient can be cast in the following analytic form,

$$\mu_{ext}^{ADA} = \frac{\pi a_{max}^2 N_t}{s^2}\left[2(s+2)(s+1) - \frac{4(s+1)s\lambda'}{(1+(s\lambda')^{-2})^{(s/2+1)}}\right.$$

$$\times \sin\left((s+2)\tan^{-1}\left(\frac{1}{s\lambda'}\right)\right) + 4(s\lambda')^2 -$$

$$\frac{4(s\lambda')^2}{(1+(s\lambda')^{-2})^{(s+1)/2}} \cos\left((s+1)\tan^{-1}\left(\frac{1}{s\lambda'}\right)\right)\Bigg], \tag{4.174}$$

where

$$\lambda' = \frac{\lambda}{4\pi m a_{max}}, \tag{4.174a}$$

and $a_{max} = s a_s$ is the most likely particle radius.

4.8.2 Modified gamma distribution for coal, fly ash and soot

Kim and Lior (1995) have given analytic expressions for extinction and scattering coefficients for coal, fly ash and soot particles. These expressions have been found to give an error of less than 10%.

Coal and char particles: The large particle approximations (3.228) and (3.229), applicable to single pulverized coal particle, when multiplied with the particle size distribution (4.115) and integrated, lead to

$$\mu_{ext} = \frac{2A\Gamma(a+3)}{N\bar{r}b^{a+3}}, \tag{4.175}$$

and

$$\mu_{sca} = \frac{a(F_1+F_2)}{2N\bar{r}^2}\frac{\Gamma(a+3)}{b^{a+3}}, \tag{4.176}$$

where

$$\bar{r} = \frac{a+1}{b}, \tag{4.177}$$

is the overall mean radius, and F_1 and F_2 are given by (3.230).

Fly ash particles: For fly ash particles, the anomalous diffraction limit is applicable. Use of the ADA, along with the size distribution (4.115), gives,

$$\mu_{ext} = \frac{A}{Na^2}\left[\frac{2\Gamma(a+3)}{b^{a-3}} - \frac{\cos\phi \sin[(a+2)\psi - \phi]\,\Gamma(a+2)}{C_1^{a-3}(C_2^2+1)^{(a+2)/2}}\right.$$

$$\left.\frac{\cos^2\phi\cos[(a+1)\psi-\phi]\,\Gamma(a+1)}{C_1^{a+3}(C_2^2+1)^{(a+1)/2}} + \frac{4\cos^2\phi\cos(2\phi)\Gamma(a+1)}{C_1^2 b^{a+1}}\right], \tag{4.178}$$

and

$$\mu_{sca} = \frac{a}{N\bar{r}^2}\left[\frac{\Gamma(a+3)}{b^{a+3}} + \frac{C_3\Gamma(a+2)}{C_4^{a+2}}\right.$$

$$+\frac{C_3^2}{2}\left(\frac{\Gamma(a+1)}{C_4^{a+1}}-\frac{\Gamma(a+1)}{b^{a+1}}\right)\Bigg], \quad (4.179)$$

where

$$C_1=\frac{4\pi}{\lambda}(n_r-1),\quad C_2=\frac{b}{C_1}+\tan\phi,\quad C_3=\frac{\lambda}{4\pi n_i},\quad C_4=b+\frac{2}{C_3}, \quad (4.180a)$$

and

$$\psi=\tan^{-1}\frac{1}{C_2}. \quad (4.180b)$$

Soot particles: For soot particles, $x \ll 1$, and the Penndorf approximation may be used to derive expressions for μ_{ext} and μ_{sca}:

$$\mu_{ext}=\frac{a}{N\bar{r}^2}\left[c_1\left(\frac{2\pi}{\lambda}\right)\frac{\Gamma(a+4)}{b^{a+4}}+c_2\left(\frac{2\pi}{\lambda}\right)^3\frac{\Gamma(a-6)}{b^{a-6}}+c_3\left(\frac{2\pi}{\lambda}\right)^4\frac{\Gamma(a+7)}{b^{a+7}}\right], \quad (4.181)$$

$$\mu_{sca}=\frac{a}{N\bar{r}^2}\left[c_1\left(\frac{2\pi}{\lambda}\right)\frac{\Gamma(a+4)}{b^{a-4}}+c_2\left(\frac{2\pi}{\lambda}\right)^3\frac{\Gamma(a+6)}{b^{a+6}}+c_4\left(\frac{2\pi}{\lambda}\right)^4\frac{\Gamma(a+7)}{b^{a-7}}\right.$$
$$\left.+c_5\left(\frac{2\pi}{\lambda}\right)^5\frac{\Gamma(a+9)}{b^{a-9}}+c_6\left(\frac{2\pi}{\lambda}\right)^7\frac{\Gamma(a-10)}{b^{a+10}}\right], \quad (4.181a)$$

where

$$c_1=\frac{24n_rn_i}{4n_r^2n_i^2+(2-n_i^2+n_r^2)^2},$$

$$c_2=4n_rn_i\left[\frac{1}{15}+\frac{5}{3}\frac{1}{16n_r^2n_i^2+(3+2(n_r^2-n_i^2))^2}\right.$$
$$\left.+\frac{6}{5}\frac{\left[7(n_r^2+n_i^2)^2+4(-5-n_i^2+n_r^2)\right]}{\left[4n_r^2n_i^2+(2-n_i^2+n_r^2)^2\right]^2}\right],$$

$$c_3=\frac{8}{3}\left[1+\frac{(1+n_r^2+n_i^2)^2-4n_r^2}{4n_r^2n_i^2+(2-n_i^2+n_r^2)^2}+\right.$$
$$\left.\frac{\left[-2-n_i^2-n_r^2+(n_r^2+n_i^2)^2\right]^2-36n_r^2n_i^2}{\left[4n_r^2n_i^2+(2-n_i^2+n_r^2)^2\right]^2}\right]$$

$$c_4=\frac{8}{3}\left[1-\frac{(1+n_r^2+n_i^2)^2-4n_r^2}{4n_r^2n_i^2+(2-n_i^2+n_r^2)^2}\right],$$

$$c_5=\frac{-16}{5}\left[\frac{\left[(n_r^2+n_i^2)^2-4\right]\left[(1+n_i^2+n_r^2)^2-4n^2\right]}{\left[4n_r^2n_i^2+(2-n_i^2+n_r^2)^2\right]^2}\right],$$

$$c_6 = \frac{-32}{3}\left[2N_r n_i\left[(1+n_r^2+n_i^2)^2 - 4n^2\right]\left[4n_r^2 n_i^2 + (2-n_i^2+n_r^2)^2\right]^2\right]. \tag{4.182}$$

In addition to the above approximations, the TDRA (3.19) was also included in the study. For a collection of particles with size distribution (4.115), it leads to

$$\mu_{ext} = \frac{2A}{Nr^r}\left[\frac{\Gamma(a+3)}{b^{a-3}} - \frac{\Gamma(a+3)}{\left[4\pi(n_r-1)\frac{G}{\lambda}\right]+b^{a-3}}\right], \tag{4.183}$$

where G is as given by (3.19a).

4.8.3 Power law distribution

The power law size distribution for scatterers, often also referred to as the fractal size distribution, has been frequently used in diverse contexts including models for biomedical tissues, interstellar dust, aerosol particles, particles in marine environments, etc.

Particles of interest in biomedical optics: For biomedical tissues, models using power law size distribution appear to have been initiated by Xu and Alfano (1995) and Gelebart et al. (1996) for predicting/extracting optical properties of a tissue. Subsequently, its efficiency and efficacy has been studied by many researchers including, Schmitt and Kumar (1998), Wang (2000), Sharma and Banerjee (2003, 2005, 2012), Passos et al. (2005), Sheppard (2007), Schneiderheinze et al. (2007), Turzhitsky et al. (2010) and many more. The model assumes a size distribution of the form:

$$\eta(a) = \eta_0 a^{3-D_f}, \tag{4.184}$$

where, the exponent D_f need not be an integer. For this reason, the distribution is also referred to as the fractal distribution. In the framework of the ADA, the following expression for the scattering coefficient has been obtained by Sharma and Banerjee (2012):

$$\mu_{sca}^{ADA} = \frac{3\eta_0}{2}\left[\frac{2}{-D_f+3}\frac{1}{d^{D_f-3}} + \frac{4\lambda^2}{\beta^2(-D_f+1)}\frac{1}{d^{-1-D_f}} + \right.$$

$$\frac{2\lambda^{-D_f+3}}{\beta^{-D_f+3}}\left(e^{-i\pi(1+D_f)/2}\Gamma(-D_f+2,-i\beta d/\lambda) + e^{i\pi(D_f-1)/2}\Gamma(-D_f+2,i\beta d/\lambda)\right.$$

$$\left.\left. +e^{i\pi(1-D_f)/2}\Gamma(-D_f+1,-i\beta d\lambda) + e^{i\pi(D_f-1)/2}\Gamma(-D_f+1,i\beta d/\lambda)\right)\right]_{d_{min}}^{d_{max}}, \tag{4.185}$$

where $\beta = 2\pi(m-1)$ and Γ is the gamma function. For integer values of D_f, simpler expressions for μ_{sca}^{ADA} are possible. For example, for $D_f = 5$,

$$\mu_{sca}^{ADA} = \frac{3\eta_0}{2}\left[\frac{1}{d^2}\left(\frac{\lambda^2}{\beta^2 d^2}(\cos\beta d/\lambda - 1) + \frac{\lambda\sin\beta d/\lambda}{\beta d} + \frac{\cos\beta d/\lambda - 2}{2}\right.\right.$$

$$\left.\left.-\frac{\lambda\beta d\sin(\beta d/\lambda)}{2}\right)\right]_{d_{min}}^{d_{max}} + \frac{3\eta_0\beta^2}{4\lambda^2}\Big[ci(\beta d_{max}/\lambda) - ci(\beta d_{min}/\lambda)\Big], \qquad (4.186)$$

where

$$ci(x,y) = -\int_x^\infty \frac{\cos(ty)}{t}dt. \qquad (4.186a)$$

For a soft tissue model d_{min} and d_{max} have been generally taken to be roughly around $0.005 - 30\mu m$ and the relative refractive index of scatterers is $m \leq 1.1$. For a wavelength $\lambda = 0.5428\mu m$, $\beta d_{max}/\lambda \gg 1$ and $\beta d_{min}/\lambda \ll 1$. When these conditions are satisfied, the contribution from the upper limit can be ignored in comparison to that from the lower limit. It is then possible to write (3.186) as,

$$\mu_{sca}^{ADA} \approx \frac{3\eta_0\beta^2}{4\lambda^2}\left[\frac{1}{4} - C + \ln(\beta d/\lambda)\right], \qquad (4.187)$$

That is $\mu_s^{ADA} \propto \lambda^{-2}$. This is consistent with the observations made by Wang (2000).

A numerical comparison of exact and ADA predicted scattering coefficients has been carried out by Sharma and Banerjee (2012) for four values of fractal dimension $D_f = 3.0, 4.0, 4.2$ and 5.0 for $0.6 < \lambda < 1.5$ μm and $0.005 < d < 30.0$ μm. The agreement is excellent.

Particles of interest in marine environments: Matciak (2012) has obtained analytic formulas for scattering coefficients in the context of marine particles assuming a power law size distribution of the form,

$$\eta(d) = C\left(\frac{d}{d_0}\right)^\alpha. \qquad (4.188)$$

The following expression for the scattering coefficient, defined in (4.100), is obtained in the framework of the ADA,

$$b(\lambda)_{ADA} = C\ S\ \frac{\Gamma(4-\alpha)}{\alpha - 3}\left(2\pi n_w\frac{n_r - 1}{\lambda}\right)^{\alpha-3} \times \left[2\left(\left(\frac{n_i}{n_r - 1}\right)^2 + 1\right)^{(\alpha-3)/2}\right.$$

$$\left.\cos\left((\alpha-3)\tan^{-1}\left(\frac{n_r - 1}{n_i}\right)\right) - \left(2\frac{n_i}{n_r - 1}\right)^{\alpha-3}\right], \qquad (4.189)$$

where,

$$S = \left\langle \int \int_{\frac{P}{d^2}} \left(\frac{r}{d}\right)^{\alpha-3} d\left(\frac{P}{d^2}\right)\right\rangle, \tag{4.190}$$

is a shape dependent factor. In the equation (4.189), $n_w = 1.33$, n_r and n_i are the particles' real and imaginary parts of the refractive index, d is the characteristic particle size, and when divided by unit size d_0, the ratio d/d_0 makes the particle size distribution dimensionless. As $\alpha \to 4$, $\Gamma(\alpha - 4) \to \infty$. Even in this case, an analytic approximation for $\Gamma(\alpha - 4)$ can be obtained using the *Mathematica* software.

For nonabsorbing particles, the expressions obtained for scattering coefficients are:

$$b(\lambda)_{ADA} = 2\alpha\ S\ \cos\left[0.5\pi(\alpha-3)\right] \frac{\Gamma(4-\alpha)}{\alpha-3}\left(2\pi n_w \frac{n_r - 1}{\lambda}\right)^{\alpha-3}, \tag{4.191}$$

when $3 < \alpha < 4$ and $4 < k < 5$; and

$$b(\lambda) = \pi\alpha\ S\ \left(2\pi n_w \frac{n_r - 1}{\lambda}\right), \tag{4.192}$$

when $\alpha = 4$. These expressions give reliable light scattering spectra of marine particles with the power law size distribution.

Particles in atmospheric aerosols: Nicholls (1984) has given an analytic expression for the extinction coefficient for an ensemble of particle size distributions:

$$\eta(a) = Ca^{-\alpha}. \tag{4.193}$$

The extinction coefficient can be expressed as,

$$\mu_{ext}^{ADA} = 4\pi C \sum_s B_s(\alpha)/\lambda^{2s}, \tag{4.194}$$

where

$$B_s(\alpha) = \frac{(-1)^{s+1}(2s+1)J^{2s}\left[a_2^{2s+3-\alpha} - a_1^{2s+3-\alpha}\right]}{(2s+2)!(2s+3-\alpha)}, \tag{4.195}$$

with a_1 and a_2 as the lower and upper limit of the integration, $J = 4\pi(m-1)$ and

$$C = \frac{N(\alpha-1)}{\left[a_1^{(1-\alpha)} - a_2^{(1-\alpha)}\right]}, \tag{4.195a}$$

with N as the number density of particles.

4.8.4 Power law distribution: Empirical formulas for interstellar extinction

Analysis of the extinction spectrum of an interstellar medium is known to give important information about the interstellar dust constituents. Reliable extinction data is available in the spectral region $0.1 \leq \lambda \leq 5$ μm for an average extinction of the Milky Way.

In astronomy and astrophysics, the extinction coefficient is defined in a slightly different manner (see, for example, Whittet 2002):

$$\frac{E_{\lambda-V}}{E_{B-V}} = \frac{A(\lambda) - A_V}{E_{B-V}} = R_V \left[\frac{A_\lambda}{A_V} - 1 \right], \tag{4.196}$$

where

$$A(\lambda) = 1.086\pi \int a^2 Q_{ext}(a)\eta(a)da, \tag{4.197}$$

is proportional to $K_{ext}(\lambda)$. Thus, K_{ext} defined earlier is proportional to $E_{\lambda-V}/E_{B-V}$. The average extinction spectrum for the Milky Way, for which $R_V = 3.05$, is shown in Figure 4.4.

The following empirical relations have also been proposed by Fitzpatrick and Masa (1988, 1990) that fit the extinction spectrum:

$$\frac{E_{\lambda-V}}{E_{B-V}} = (C_1 + C_2 x') + C_3 D(x') + C_4 F(x'), \tag{4.198}$$

where $x' = 1/\lambda$,

$$F(x') = 0.5392(x' - 5.9)^2 + 0.0564(x' - 5.9)^3 \quad \text{for} \quad x' > 5.9 \ \mu m^{-1}, \tag{4.199a}$$

and

$$F(x') = 0 \quad \text{for} \quad x' \leq 5.9 \ \mu m^{-1}. \tag{4.199b}$$

The Drude profile, $D(x')$, is defined as

$$D = \frac{x^2}{(x - x_0)^2 + x^2\gamma^2}, \tag{4.200}$$

where x_0 is the position and γ is the width of the bump. The extinction curve may then be fitted with six parameters $C_1, C_2, C_3, C_4, \gamma$ and x_0.

Another attempt to fit the extinction spectrum is by Cardelli et al. (1989). The general structure of the formula is:

$$\langle A(\lambda)/A(V) \rangle = a(x') + b(x')/R_V. \tag{4.201}$$

Specific formulas in different spectral regions are:

(i) In the infrared and optical region; $0.3\ \mu m^{-1} \leq x' \leq 1.1\ \mu m^{-1}$,

$$a(x') = 0.574(x')^{-1.61} \text{ and } b(x') = -0.527(x')^{1.61}. \tag{4.202}$$

(ii) In the optical-near infrared region; $1.1\ \mu m^{-1} \leq x' \leq 3.3\ \mu m^{-1}$ and $y = (x - 1.82)$,

$$a(x') = 1 + 0.17699y - 0.50477y^2 - 0.02427y^3 + 0.72085y^4 + 0.01979y^5$$
$$-0.77530y^6 + 0.32999y^7, \tag{4.203a}$$

and

$$b(x') = 1.41338y + 2.28305y^2 + 1.07233y^3 - 5.58434y^4 - 0.62251y^5$$
$$+5.30260y^6 - 2.09002y^7. \tag{4.203b}$$

(iii) In the ultraviolet and the far ultraviolet region; $3.3\ \mu m^{-1} \leq x \leq 8\ \mu m^{-1}$,

$$a(x') = 1.752 - 0.316x' - \frac{0.104}{(x' - 4.67)^2 + 0.341]} + F_a(x'), \tag{4.204a}$$

and

$$b(x') = -3.09 + 1.825x' - \frac{1.206}{(x' - 4.62)^2 + 0.263]} + F_b(x'), \tag{4.204b}$$

where

$$F_a(x') = -0.04473(x' - 5.9)^2 - 0.009779(x' - 5.9)^3 \text{ for } 8 \geq x' \geq 5.9, \tag{4.205a}$$

$$F_b(x') = 0.2130(x' - 5.9)^2 + 0.1207(x' - 5.9)^3 \text{ for } 8 \geq x' \geq 5.9, \tag{4.205b}$$

$$F_a(x') = F_b(x') = 0 \quad \text{for} \quad x < 5.9. \tag{4.205c}$$

(iv) Finally, in the far-ultraviolet region; $8\ \mu m^{-1} \leq x' \leq 10\ \mu m^{-1}$,

$$a(x') = -1.073 - 0.628(x' - 8) + 0.137(x' - 8)^2 - 0.070(x' - 8)^3, \tag{4.206a}$$

and

$$b(x') = 13.670 + 4.257(x' - 8) - 0.420(x' - 8)^2 + 0.374(x' - 8)^3. \tag{4.206b}$$

The formulas are essentially empirical in nature.

A classic model of interstellar dust was proposed long ago by Mathis, Rumpl and Nordsieck (MRN) (1977). The same basic model is being used even today, albeit with some modifications. The MRN model assumes two separate populations of spherical bare silicate and graphite grains as constituents of interstellar dust. The size distribution of both types of grains is taken to be:

$$f(a) \propto a^{-3.5}, \tag{4.207}$$

where a is the radius of the spherical grains in the limits $a_0 \leq a \leq a_m$. The numerical values for a_0 and a_m are:

$$\text{Graphite grains: } a_0 \sim 0.005\ \mu m \quad, a_m \sim 0.25\ \mu m, \tag{4.208a}$$

$$\text{Silicate grains: } a_0 \sim 0.005\ \mu m \quad, a_m \sim 0.25\ \mu m. \tag{4.208b}$$

Graphite is a highly anisotropic material. Therefore, its refractive index depends on orientation of the electric field relative to the crystal axis. Effect of this anisotropy is taken into account by resorting to an approximation known as the "$\frac{1}{3} - \frac{2}{3}$" approximation (Draine and Malhotra 1993). This approximation assumes graphite grains as a mixture of isotropic spheres, $\frac{1}{3}$ of which have refractive index $m = m_{\parallel}$ and $\frac{2}{3}$ have the refractive index $m = m_{\perp}$. This modification essentially makes the MRN model a three-component model.

Roy et al. (2009, 2010) computed the extinction spectrum of the three components of the dust separately using Mie theory and parametrized the resulting extinction spectrum separately as a function of size and frequency. This allows quick and efficient assessment of the extinction contributions of the silicate and graphite contributions separately. This can be of immense help in analyzing any extinction spectrum from other stars and galaxies. The analytic expressions so obtained for the extinction coefficient in the region $1000 \leq \lambda \leq 22,500\ \AA$ have a general structure:

$$K_{ext} = CNa_{min}^{5/2}\Big[\phi(a_0, \nu) + \psi(a_m, \nu)\Big]. \tag{4.209}$$

The functions ϕ and ψ have forms which change in various frequency subintervals. The formulas given below are valid for $0.002 \leq a_0 \leq 0.005\ \mu m$ and $0.15 \leq a_m \leq 0.25\ \mu m$ for both variants of graphite. The corresponding region for silicate is $0.004 \leq a_0 \leq 0.006\ \mu m$ and $0.2 \leq a_m \leq 0.4\ \mu m$. The following relationships were obtained:

Homogeneous graphite grains with refractive index $m = m_{\perp}$

(i) $1000 \leq \lambda\ 1460\AA$ ($Far-ultraviolet\ region\ I$)

$$K_{ext} = C(a_0)^{5/2}\Big[\nu^2\left(0.259 + 20.3073(\nu - 0.8485)^4\right) - (\nu a_0)^{1/2}\Big(0.267\nu -$$

$$0.16048 + 71.15(\nu - 0.8428)^4\Big) - a_0\nu^{5/2}\Big(0.0458\nu + 0.0164$$

$$+14.827(\nu - 0.8428)^2(\nu - 0.67905)\Big) - 0.0488\left(\frac{1}{a_m^{1/2}} + \frac{0.1551}{a_m}\right)\Bigg]. \quad (4.210)$$

(ii) $1460 \leq \lambda \leq 1900$ Å (*Far − ultraviolet region II*)

$$K_{ext} = C(a_0)^{5/2}\Bigg[1.97754 - \frac{2.42703}{\nu} + \frac{0.79483}{\nu^2} - \frac{a_0^{1/2}}{\nu^2}\Big(0.2164 - 0.27812\nu +$$

$$\frac{0.43347}{\nu^{1/4}}\left(1.0 - \frac{0.6856}{\nu}\right) \times \left(1.0 - \frac{0.5263}{\nu}\right)\Big) - \frac{a_0}{2\nu^2}\Big(\left(1.0 - \frac{0.606}{\nu}\right)^2 +$$

$$\frac{0.03173}{\nu}\left(1.0 - \frac{0.6262}{\nu}\right) - 46.52\nu(\nu - 0.6856)(\nu - 0.5263)(\nu - 0.606)\Big) -$$

$$0.0488\left(\frac{1}{a_m^{1/2}} + \frac{0.1551}{a_m}\right)\left(0.8667 + 24.1(\nu - 0.6112)^2\right)\Bigg]. \quad (4.211)$$

(iii) $1900 \leq \lambda \leq 2400$Å (*Ultraviolet region I*)

$$K_{ext} = Ca_0^{5/2}\Bigg[\frac{1}{2.7 + 490(\nu - 0.4654)^2} + 100|\nu - 0.4654|^3 - \nu a_0^{1/2}$$

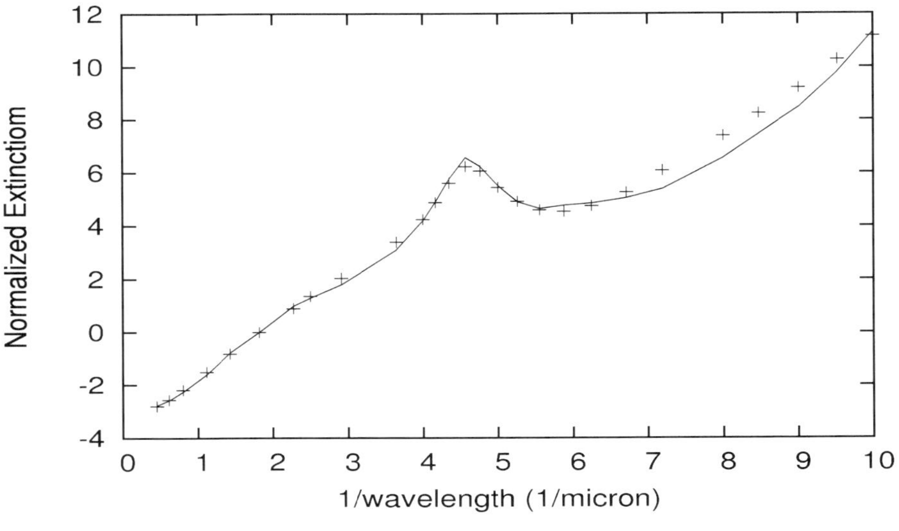

FIGURE 4.4: A comparison of observed average interstellar extinction data (solid line) and that obtained using the formulas of Roy et al. (2009, 2010) (points). This figure was first published in Roy et al. (2010).

$$\times\left(\frac{1}{1.1058+6.5985|\nu-0.4736|}+11307(\nu-0.4673)^4-258.26\nu(\nu-0.4795)^2\right)$$

$$-a_0\left(\frac{0.0813}{1.0+6.9\left(1.0-\frac{0.4566}{\nu}\right)+76.187\nu|\nu-0.452|}\right)$$

$$-0.05511\left(\frac{1}{a_m^{1/2}}+\frac{0.2398}{a_m}\right)(1.9503-1.712\nu+0.583|\nu-0.452|)\Bigg]. \quad (4.212)$$

(iv) $2400 \leq \lambda \leq 3800\text{Å}$ (*Ultraviolet region II*)

$$K_{ext}=Ca_0^{5/2}\Bigg[\nu^{1/2}\Big(0.1161+0.60\nu-1.9684\nu^{1/4}(0.4-\nu)^{1/2}(nu-0.25)\Big)$$

$$-a_0^{1/2}\nu^2\left(0.461+1.7567|\nu-0.325|+180.6(\nu-0.325)^3\right)-a_0\nu^3$$

$$\times\left(20664(\nu-0.3244)^4+\frac{1.3019\nu}{1.0+304.9\left(1.0-\frac{0.3229}{\nu}\right)^2}\right)$$

$$-0.0507\left(\frac{1}{a_m^{1/2}}+\frac{0.278}{a_m}\right)\Bigg]. \quad (4.213)$$

(v) $4000 \leq \lambda \leq 8000$ Å (*Visible region*)

$$K_{ext}=Ca_0^{5/2}\Bigg[-0.02738\nu a_0^{1/2}\Big(1+2.5647a_0^{1/2}\Big)\left(1+3.4328\nu^{1/2}\right)+\nu\Big(0.37707$$

$$-0.02357a_m^{1/2}+0.045a_m\Big)+a_m\left(0.002808-0.12\nu^2\right)-0.02845/a_m^{3/2}\Bigg]. \quad (4.214)$$

(vi) $8000 \leq \lambda \leq 12500$ Å (*Infrared region I*)

$$K_{ext}=Ca_0^{5/2}\Bigg[-0.16941\nu^{3/2}\left(2.6493a_0+a_0^{1/2}\right)+4.2\Big(\nu-0.08)(0.125-\nu)$$

$$\times(a_m-1.5)(2.77-a_m)+\nu\Big(0.33934+0.1406a_m-0.035086a_m^2\Big)$$

$$+0.05636/a_m^{1/2}-0.237187/a_m^{3/2}+0.17054/a_m^{5/2}\Bigg]. \quad (4.215)$$

(vii) $12500 \leq \lambda \leq 22500$ Å (*Infrared region II*)

$$K_{ext} = Ca_0^{5/2}\Bigg[-a_0^{1/2}\nu^{3/2}\left(0.4083a_0^{1/2}+0.1798\right)+\nu\Big(1.12315-\frac{1.104}{a_m}$$

$$-0.043a_m\Big)+5\left(\nu-0.044\right)\left(0.08-\nu\right)\left(a_m-2.45\right)\left(a_m-1\right)-\frac{0.12273}{a_m}+\frac{0.133}{a_m^{3/2}}\Bigg]. \tag{4.216}$$

Homogeneous graphite grains with refractive index $m = m_{\|}$

(i) $1000 \leq \lambda \leq 1120$ Å (*Far − ultraviolet region I*)

$$K_{ext} = Ca_0^{5/2}\Bigg[\nu^{3/2}\left(0.3473-6.284(1-\nu)^2\right)-a_0^{1/2}\nu^{3/2}\Big(2.3047\nu-$$

$$2.0047+1.38\left(1-\frac{0.901}{\nu}\right)\left(\frac{1}{\nu}-1\right)^{1/2}\Big)$$

$$a_0\Big(5.55(\frac{1}{\nu}-1)^{1/2}\left(1-\frac{0.901}{\nu}\right)|1-\frac{0.9524}{\nu}|^{1/2}+$$

$$\frac{0.2979}{\nu^{1/2}}-\frac{11.65}{\nu^{1/2}}(1-\nu)^2\Big)-\left(\frac{0.046}{a_m^{1/2}}+\frac{0.011}{a_m}\right)\Bigg]. \tag{4.217}$$

(ii) $1120 \leq \lambda \leq 1500$ Å (*Far − ultraviolet region II*)

$$K_{ext} = Ca_0^{5/2}0.06685-0.06056\nu+0.277\nu^2-a_0^{1/2}\left(0.15482\nu-0.09754\right)$$

$$-a_0\left(0.75484\nu-0.495\right)-0.018\left(\frac{2}{a_m^{1/2}}+\frac{1}{a_m}\right)\Bigg]. \tag{4.218}$$

(iii) $1500 \leq \lambda \leq 2400$ Å (*Far − ultraviolet region III*)

$$K_{ext} = Ca_0^{5/2}\Bigg[\frac{1}{\nu^{1/2}}\left(0.864+\frac{0.79}{\nu}(\nu-0.4675)^2-7.16(\nu-0.49)^4\right)-\frac{a_0^{1/2}}{\nu^{5/2}}$$

$$\times\left(0.0065+\frac{0.06575}{\nu}(\nu-0.4922)^2-1.76(\nu-0.5112)^4\right)-a_0\Big(0.0054+$$

$$\frac{0.24|\nu-0.5263|}{1.0+118(\nu-0.5405)^2}\Big)-\frac{0.03913+0.07925|\nu-0.5311|}{a_m^{1/2}}-$$

$$\left. \frac{0.008 - 0.2116(\nu - 0.5486)^2}{\nu a_m} \right]. \tag{4.219}$$

(iv) $2400 \leq \lambda \leq 4000$ Å (*Ultraviolet region*)

$$K_{ext} = Ca_0^{5/2} \left[0.1495 - \frac{0.0847}{\nu} |\nu - 0.3675| - a_0^{1/2} \Big(\nu(0.1915 - 0.817|\nu - 0.3719|) \right.$$

$$-33.0|\nu - 0.3675|(0.4167 - \nu)(\nu - 0.25)^2 \Big) a_0 \frac{0.00356}{(1 + 15.017|\nu - 0.4031|)^2}$$

$$\left. - \frac{0.26}{a_m^{1/2}} \nu (1.0 - 1.36\nu) - \frac{0.00436 + 0.0122\nu|\nu - 0.3542|}{a_m \nu} \right]. \tag{4.220}$$

(v) $4000 \leq \lambda \leq 8000$ Å (*Visible region*)

$$K_{ext} = Ca_0^{5/2} \left[-1.10 a_0^{1/2} \nu^3 + 3.6927 \left(\nu^{1/2} - 0.3535 \right) \left(0.5 - \nu^{1/2} \right) \right.$$

$$\left. \left(a_m - \frac{1.4535}{a_m} \right) \times \left(\frac{1}{a_m^{1/2}} - \frac{0.1737}{a_m^{3/2}} \right) - \frac{0.07946}{a_m^{1/2}} + \frac{0.02357}{a_m^{1/2}} - \frac{0.00939}{a_m^{3/2}} \right]. \tag{4.221}$$

(vi) $8000 \leq \lambda \leq 12500$ Å (*Infrared region I*)

$$K_{ext} = Ca_0^{5/2} \left[\nu \left(\frac{1.5635}{a_m^{1/2}} - \frac{1.9781}{a_m^{3/2}} \right) - \frac{0.18925}{a_m} + 0.00713 \right.$$

$$\left. + \frac{0.236022}{a_m^2} + 8.0 (\nu - 0.08) (\nu - 0.125) (a_m - 1) (2.5 - a_m) \right]. \tag{4.222}$$

(vii) $12500 \leq \lambda \leq 22500$ Å (*Infrared region II*)

$$K_{ext} = Ca_0^{5/2} \nu^2 \left[-0.1 a_0^{1/2} + \nu^3 a_m \left(32.95 + 49.571 a_m^3 \right) \right.$$

$$\left. + a_m \left(0.099423 - 0.004578 a_m^2 + 0.002301 a_m^3 \right) \right]. \tag{4.223}$$

Homogeneous silicates grains

(i) $1000 \leq \lambda \leq 1460$Å (*Far − ultraviolet region I*)

$$K_{ext} = Ca_0^{5/2}\Bigg[0.11668 - 0.02745\nu + 0.17\nu^2 - a_0^{1/2}(0.3542\nu - 0.2110)$$

$$a_0(0.6677\nu - 0.3733) + \left(2a_m(1-\nu)(\nu - 0.6856) - \frac{0.05}{a_m^{1/2}} - \frac{0.0062}{a_m}\right)\Bigg]. \quad (4.224)$$

(ii) $1400 \leq \lambda \leq 2000$ Å $(Far - ultraviolet\ region\ II)$

$$K_{ext} = Ca_0^{5/2}\Bigg[0.29168\nu - 0.02268 - a_0^{1/2}(0.0634\nu - 0.01623)$$

$$-a_0(0.64634\nu - 0.3478) - \left(\frac{0.05}{a_m^{1/2}} + \frac{0.0062}{a_m}\right)\Bigg]. \quad (4.225)$$

(iii) $2000 \leq \lambda \leq 4000$ Å $(Ultraviolet\ region)$

$$K_{ext} = Ca_0^{5/2}\Bigg[-0.02883 + 0.197\nu + \frac{0.00669}{\nu} + 0.006567a_m\nu - \frac{0.0151}{a_m^2\nu}$$

$$-\frac{0.016417}{a_m^3}\left(\frac{1 - 4|1 - \frac{0.435}{\nu}|}{\left(1.0 + 10|1 - \frac{0.435}{\nu}|\right)^2}\right)\Bigg]. \quad (4.226)$$

(iv) $4000 \leq \lambda \leq 8000$ Å $(Visible\ region)$

$$K_{ext} = Ca_0^{5/2}\Bigg[a_m\left(0.02884 - 0.02613a_m + a_m^2\right) - \nu a_m^{3/2}\Big(0.2673 - 0.2258a_m$$

$$+0.0365a_m^2\Big) + \nu^2 a_m^2\left(2.73017 - 3.06825a_m^{1/2} + 0.8417a_m\right)\Bigg]. \quad (4.227)$$

(v) $8000 \leq \lambda \leq 22500$ Å $(Infrared\ region)$

$$K = Ca_0^{5/2}\Bigg[0.0004 - 0.0016\nu^{1/2} - 0.007871a_0^{1/2}\nu + 0.01835a_m^{1/2}\nu$$

$$a_m^{3/2}\nu^2\left(0.2323 - 1.8104\nu a_m + 2.0856\nu^2 a_m^2\right)\Bigg]. \quad (4.228)$$

In these formulas $C = 2\pi\mathcal{N}$ with $\mathcal{N} = 2.2 \times 10^8$ and a and λ are in the units of $10^{-5}cm$. A fit using the above formulas is shown in Figure 4.4.

4.9 Scattering by nonspherical polydispersions

The effect of nonsphericity in an ensemble of Rayleigh particles has not been found to be significant. Paramonov (1994) found that the light scattered by randomly oriented ellipsoids is almost indistinguishable from that scattered by a collection of equal volume spheres (Paramonov 1994). Further work by Min et al. (2006) has shown equivalence between the absorption cross section of an ensemble of particles of any shape, with the absorption cross section of an ensemble of spheroidal particles.

Paramonov (2012), within the realm of the RGA, has shown that randomly oriented ellipsoidal particles in an isotropic ensemble are optically equivalent to a polydispersion of randomly oriented spheroidal particles and a polydispersion of spherical particles. This was found to be true for scattering and absorption cross sections in the framework of the ADA too.

Ensembles of nonspherical particles are generally replaced by equal volume or equal surface area spheres. However, significantly improved equivalence has been obtained if the ensemble of nonspherical particles is substituted by an ensemble of spherical particles with the same volume/surface area ratio. Full analytic proof of the applicability of this approach has been presented by Shepelevich et al. (2001).

Single scattering properties of dust aerosols and their effect in radiative flux calculations for spherical and spheroidal particles has been studied by many researchers. Fu et al. (2009) studied this for lognormal size distributions. The equivalent sphere was defined as one giving the same (volume/projected area). The results showed that the errors in extinction efficiency and albedo in approximating spheroids with spheres is less than 1%. The errors in the asymmetry parameter were less than 2%. The phase function obtained by approximating spheroids by spheres was found to be better than that predicted by HGPF in the angular range 0-90 deg.

4.10 Effective phase function

Apart from single scattering phase function, higher-order scattering phase functions are also of interest from the point of view of diagnosing/characterising the scattering media.

In this context, it has been shown by Turcu and coworkers (Turcu et al.

2006, 2008a, 2008b and Turcu and Kirillin 2009) that, if

$$p(\mu) = \frac{1}{4\pi} \sum_{l=0}^{\infty} (2l+1)\, a_l \; P_l(\mu), \tag{4.229}$$

is a forward peaked single scattering phase function, then the n-th order scattering phase function can be expressed as,

$$p_n(\mu) = \frac{1}{4\pi} \sum_{l=0}^{\infty} (2l+1)\, (a_l)^n \; P_l(\mu). \tag{4.230}$$

That is, the n-th order scattering phase function can be obtained by replacing the expansion coefficient in the single scattering phase function by the n-th power of the coefficient of the corresponding single particle phase function.

One may also define an effective phase function which describes the angular spreading of the scattered flux at any optical depth τ. It can be written as,

$$p_{eff}(\tau, \mu) = \frac{\Phi_{sca}}{(1 - e^{-\tau})}, \tag{4.231}$$

where Φ_{sca} is,

$$\Phi_{sca}(\tau, \mu) = \Phi(\tau, \mu) - \Phi_0(\tau, \mu), \tag{4.232)}$$

Φ_0 is the reduced incident flux and $\Phi(\tau, \mu)$ is the incident flux.

It can be argued that for quasi-ballistic photons, the scattered flux Φ_{sca} is

$$\Phi_{sca}(\tau, \mu) = \sum_{n=1}^{\infty} \Phi_n(\tau, \mu), \tag{4.233}$$

with

$$\Phi_n = \frac{\tau^n}{n!} e^{-\tau} p_n(\mu), \tag{4.234}$$

as the relation between n-th order scattered flux and the n-th order phase function.

For the special case of the Henyey–Greenstein phase function,

$$p_{HGPF}(\mu) = \frac{1}{4\pi} \frac{1 - g^2}{(1 - 2g\mu + g^2)3/2} = \frac{1}{4\pi} \sum_{l=0}^{\infty} (2l+1)\, g^l P_l(\mu), \tag{4.235}$$

the n-th order phase function, according to the aforesaid prescription, becomes

$$p_n^{HGPF}(\mu) = \frac{1}{4\pi} \sum_{l=0}^{\infty} (2l+1)\, (g^l)^n P_l(\mu) = \frac{1}{4\pi} \frac{1 - g^{2n}}{(1 - 2\mu g^n + g^{2n})^{3/2}}. \tag{4.236}$$

Therefore, the effective phase function (4.231) becomes,

$$p_{eff} = \sum_{n=1}^{\infty} \frac{\tau^n}{n!} \frac{p_n^{HGPF}(\mu)}{e^\tau - 1} = \frac{1}{4\pi} \sum_{l=0}^{\infty} (2l+1) \frac{(e^{\tau g^l} - 1)}{(e^\tau - 1)} P_l(\mu). \qquad (4.237)$$

It is not easy to extract the pertinent parameters from the experimental data using this formula. Thus, two approximate expressions have been employed for the effective phase function. The first is an n-th order sum:

$$p_{eff}(\tau, \mu, g) = \frac{\sum_{k=1}^{n} p_k^{HGPF}(\mu, g) \frac{\tau^k}{k!}}{\sum_{k=1}^{n} \frac{\tau^k}{k!}}. \qquad (4.238)$$

The accuracy of (4.238) can be increased by increasing the number of terms and also by decreasing the angular range around the forward direction. The second approximation is

$$p_{eff}(\tau, \mu, g) = \frac{1}{4\pi} \frac{1 - g^{2G(\tau)}}{[1 - 2\mu g^{2G(\tau)} + g^{2G(\tau)}]^{3/2}}, \qquad (4.239)$$

where,

$$G(\tau) = \frac{(\tau - 1)e^\tau + 1}{(e^\tau - \tau - 1)}. \qquad (4.240)$$

This is a compact formula and has the advantage of simplicity. It has been successfully used to describe the small angle scattering from blood (Turcu et al. 2006). Pfeiffer and Chapman (2008) have extended this treatment to two-term phase functions too.

An alternative expression for the effective phase function, in terms of single scattering albedo and the n-th order scattering phase function, has been obtained by Piskozub and McKee (2011):

$$p_{eff}(\theta) = \frac{p_1(\theta) + \omega_0 p_2(\theta) + \ldots\ldots + \omega_0^{n-1} p_n(\theta)}{1 + \omega_0 + \ldots\ldots + \omega_0^{n-1}}, \qquad (4.241)$$

where $p_n(\theta)$ is the n-th order phase function. Piskozub and McKee (2011) have given a prescription for calculation of the n-th order phase function too. The validity of this prescription has been verified by comparing it with Pfeiffer and Chapman's (2008) results obtained by using the HGPF in the Monte Carlo simulations.

4.11 Relation between light scattering reflectance and the phase function

An experimental setup frequently employed to study the propagation of light in turbid media consists of a narrow beam of unpolarized light vertically incident on a flat surface of the scattering medium. The reflected light intensity is measured at a source detector separation ρ. The scattering medium is assumed to be isotropic. The measurement is called spatially resolved reflectance and can be analysed to characterise the medium.

The further away a photon is detected, the larger the probability is that it has penetrated deeper in the medium and undergone a larger number of scatterings. As a consequence, the reflectance with small source detector separation corresponds to shallow penetration. In other words, it corresponds to photons that have undergone subdiffusive scattering.

Bevilacqua and Depeursinge (1999) have examined the relationship between the reflectance at small source detector separation and the scattering phase function. They demonstrated that for $0.5 < \rho\mu'_{sca} < 5$, a parameter γ can be expressed as

$$\gamma = \frac{1 - g_2}{1 - g_1}, \tag{4.242}$$

which represents the relative contribution of near backward scattering in the phase function. In addition to μ_{sca}, this parameter is found to serve as a useful quantifier of subdiffusive reflectance. The parameters g_1 and g_2 are the first two moments of the phase function. The n-th moment of the phase function is defined as

$$g_n = 2\pi \int_0^\pi P_n(\cos\theta) p(\cos\theta) \sin\theta d\theta. \tag{4.243}$$

For any combination of g_1, g_2 and μ_{sca}, which gives same μ'_{sca} and γ, optical measurements of reflectance will give equivalent results. The usefulness of the parameter γ in the analysis of reflectance has been studied by many researchers including Chamot et al. 2010, Calabro and Bigio 2014, and Bodenschatz et al. (2016).

An analytic formula for describing light reflectance at an arbitrary source detector separation from a forward peaked scattering media has been given by Xu (2016). It gives dependence of light reflectance on the phase function in a closed form. Formally, the total light reflectance in this analytic expression is expressed as:

$$I(\mathbf{q}) = I(\mathbf{q})_{snake} + I(\mathbf{q})_{diffuse} + \mu_{back}/2\mu_{ext} \qquad q < q_c, \tag{4.244}$$

and

$$I(\mathbf{q}) = I(\mathbf{q})_{SAA}, \qquad q > q_c, \tag{4.245}$$

where $q_c = 2\pi\beta$ with $\beta = \mu_{abs} + \mu'_{sca}$. $I(q)_{SAA}$ is the contribution corresponding to photons with only one large angle scattering. The subscript SAA stands for small angle approximation. The contribution of snake photons is denoted by $I(\mathbf{q})_{snake}$ and the diffuse photon contribution is denoted by $I(\mathbf{q})_{diffuse}$. The analytic expressions for each of these contributions have been given by Xu (2016).

Appendix

The form factors shown in this appendix have been taken from Shapovalov (2014) with permission from the author and the publisher.

TABLE A1: Form factors for some particles with axis of symmetry.

Shape	**Form factor**($k_4 = 0$)	**Form factor**($k_4 \neq 0$)
Sphere	$\frac{3j_1(k_3R)}{k_3R}$	$\frac{3j_1(qR)}{qR}$
Half sphere	$\frac{3(j_1(k_3R)+ih_1(k_3R))}{k_3R}$	-—
Spheroid	$\frac{3j_1(k_3c)}{k_3c}$	$\frac{3j_1\left(\sqrt{(k_4R)^2+(k_3c)^2}\right)}{\sqrt{(k_4R)^2+(k_3c)^2}}$
Cylinder	$j_0\left(k_3\frac{H}{2}\right)$	$\frac{2J_1(k_4R)}{k_4R}j_0\left(k_3\frac{H}{2}\right)$
Inf. cylinder ($\theta = \pi/2$)	—	$\frac{2J_1(k_4R)}{k_4R}$
Cone	$\frac{6}{k_3H}(h_0(k_3H) - j_1(k_3H)+$ $i(1 - h_1(k_3H - j_0(k_3H)))$	provided $k_3H = k_4R = w$ $2\frac{\exp(iw)}{w}[\tilde{q} + i\tilde{p}]$, $\tilde{q} = \cos wJ_1(w) + \sin wJ_2(w)$ $\tilde{p} = \cos wJ_2(w) - \sin wJ_1(w)$
Paraboloid of revolution	$\frac{2}{k_3H}[h_0(k_3H)$ $+i(1 - j_0(k_3H))]$	-
Torus	$\frac{2J_1(k_3a)}{k_3a}$	provided $k_4R < 1$ $\frac{8J_1(k_4(R+a))J_1(k_4(R-a))J_1(k_3a)}{k_3k_4^2a(R+a)(R-a)}$
Cassini oval based body $a > c$	$\frac{2J_1(k_3c^2/2a}{k_3c^2/2a}$	—
Cassini oval based body $0 < a \leq c$	—	—

TABLE A2: Form factors for randomly oriented particles.

Particle of arbitrary shape	$\langle \Phi^2(\theta) \rangle = \frac{1}{4\pi} \int_\Omega \vert\Phi\vert^2 d\Omega$, Φ is a form factor of the particle
Particle with axis of symmetry	$\frac{1}{4\pi} \int_0^{2\pi} \int_0^{\pi} \vert\Phi\vert^2 \sin\beta \, d\beta \, d\varphi$,
Spherical electron distribution in the polyatomic gases	$\left(\frac{3j_1(qR)}{qR}\right)^2 \frac{1}{N^2} \sum_{i=1}^{N} \sum_{j=1}^{N} \frac{\sin(qr_{ij})}{qr_{ij}}$, where r_{ij} is the distance between the i-th and j-th electron
Two spheres in contact	$\left(\frac{3j_1(qR)}{qR}\right)^2 \frac{1}{4}\left(2 + \frac{\sin(2qR)}{2qR}\right)$
Cylinder	$\int_0^{\pi/2} \left[\frac{2J_1(qR\sin\beta)}{qR\sin\beta} j_0\left(\frac{qH}{2}\cos\beta\right)\right]^2 \sin\beta d\beta$
Thin rod of length L	$\frac{1}{Z}\int_0^{2Z} j_0(w)dw - [j_0(Z)]^2$, where $Z = qL$
Thin disk of radius R	$\frac{2}{Z^2}\left(1 - \frac{J_1(2Z)}{Z}\right)$, where $Z = 2qR$
Assembly of infinitely long cylinders	$\left(\frac{2J_1(qR)}{qR}\right)^2 \frac{1}{N^2} \sum_{i=1}^{N} \sum_{j=1}^{N} J_0(qr_{ij})$,
Two infinitely long cylinders in contact	$\left(\frac{2J_1(qR)}{qR}\right)^2 \frac{1}{4}(2 + 2J_0(2qR))$
Seven infinitely long cylinders	$\left(\frac{2J_1(qR)}{qR}\right)^2 \frac{1}{49}((7 + 24J_0(x) + 6J_0(2x)$ $+12J_0(\sqrt{3}x))$, where $x = qr$, r is a radius or side of the hexagon

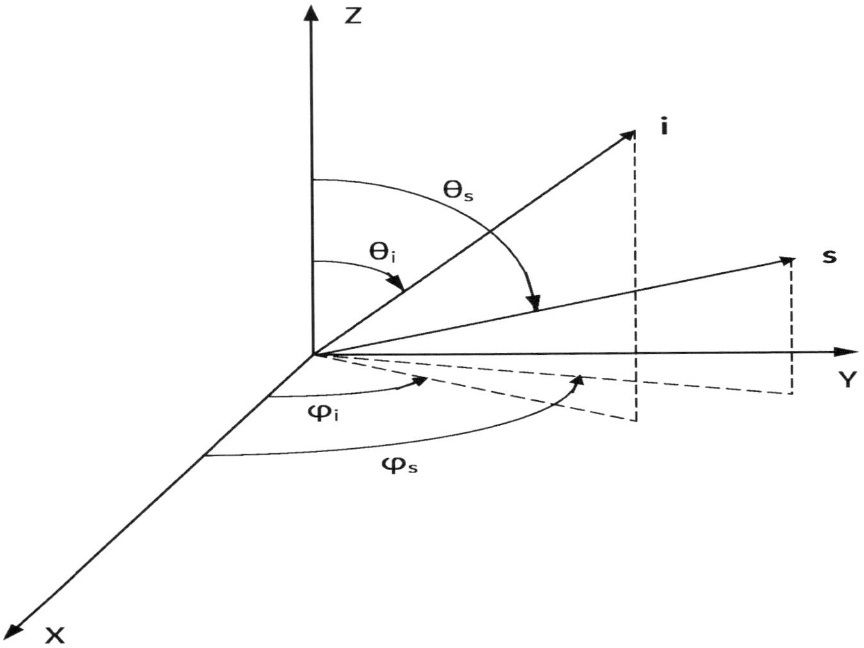

FIGURE A1: Coordinate system depicting angles specified in the Appendix.

If the incident light is expressed in terms of θ_i and φ_i and scattered light in terms of θ_s and φ_s, then,

$$k_1 = k(\sin\theta_i \cos\varphi_i - \sin\theta_s \cos\varphi_s),$$

$$k_2 = k(\sin\theta_i \sin\varphi_i - \sin\theta_s \sin\varphi_s),$$

$$k_3 = k(\cos\theta_i - \cos\theta_s),$$

$$k_4 = \sqrt{k_1^2 + k_2^2},$$

$$q = \sqrt{k_1^2 + k_2^2 + k_3^2},$$

and β is the angle between the $z-$ axis and the vector $\mathbf{q}$. R is radius of the cylinder, sphere, torus and base of the cone. $R = a = b$ for spheroid. H is the height for a cylinder, and cone and a is thickness for the torus. h_0 and h_1 are spherical Struve functions of order zero and one, respectively.

Bibliography

[1] Aas, E., 1984, Some aspects of light scattering by marine particles. *Institute of Geophysics*, Oslo.

[2] Ackerman, S. A., and Stephens, G. L., 1987. The absorption of solar radiation by cloud droplets: An application of anomalous diffraction theory. *J. Atmos. Sci.*, 44, 1574–1588.

[3] Acquista, C., 1976. Light scattering by tenuous particles: A generalization of Rayleigh–Gans–Rokard approximation. *Appl. Opt.*, 15, 2932–2936.

[4] Aden, A. L., 1951. Electromagnetic scattering from spheres with sizes comparable to the wavelength of light. *J. Appl. Phys.*, 22, 601–605.

[5] Aden, A. L., and Kerker, M., 1951. Scattering of electromagnetic waves from two concentric spheres. *J. Appl. Phys.*, 22, 1242–1246.

[6] Adey, A. W., 1956. Scattering of electromagnetic waves by a coaxial cylinder. *Can. J. Phys.*, 35, 510–520.

[7] Agrawal, Y. C., and Mikkelsen, O. A., 2009. Empirical forward scattering phase functions from 0.08 to 16 deg for randomly shaped terrigenous $1 - 21$ μm sediment grains. *Opt. Express*, 17, 8805–8814.

[8] Alam, M., and Massoud, Y., 2006. A close form analytical model for single nanoshells. *IEEE Trans. Nanotechnology*, 265–272.

[9] Allen, E., 1974. Expansion of the Mie phase function in Legendre polynomials. *Appl. Opt.*, 13, 2752–2753.

[10] Alvarez-Estrada, R. F., Calvo, M. L., and Juncos del Egido, P., 1980. Scattering of TM waves by dielectric fibres. Iterative and eikonal solutions. *Opt. Acta*, 27,1367-1378.

[11] Ambrosio, L. A., 2015. Approximations to the Mie scattering coefficients for plasmonic and negative-index Rayleigh scatterers. *IEEE Xplore*, published in IOMC, 1–5.

[12] Aragon, S. R., and Elwenspoek, M., 1982. Mie scattering from thin spherical bubbles. *J. Chem. Phys.*, 77, 3406–3413.

[13] Arnush, D., 1964. Electromagnetic scattering from a spherical nonuniform medium. Part I: General theory. *IEEE Trans. Antenn. Propagat.*, A.P-12, 86–90.

[14] Asano, S., and Yamamoto, G., 1975. Light scattering by a spheroidal particle. *Appl. Opt.*, 14, 29–49.

[15] Aspnes, D. E., 1982. Local field effects and effective medium theory: A microscopic perspective. *Am. J. Phys.*, 50, 704–709.

[16] Babenko, V. A., Astafyeva, L. G., and Kuzmin, V. N., 2003. Electromagnetic scattering in disperse media, *Springer/Praxis*, Chichister, UK.

[17] Bai, L., Wu, Z., Tang, S., Li, M., Xie, P., and Wang, S., 2011. Study on phase function in Monte Carlo transmission characteristics of polydisperse aerosols. *Opt. Engg.*, 50, 16002-1-8.

[18] Barabas, M., 1987. Scattering of plane waves by a radially stratified tilted cylinder. *J. Opt. Soc. Am. A,* 4, 2240–2248.

[19] Barber, P. W., and Wang, D. S., 1978. Rayleigh–Gans–Debye applicability to scattering by nonspherical particles. *Appl. Opt.*, 17, 797–803.

[20] Bascom, P. A. J., and Cobbold, R. S. C., 1995. On the fractal packing approach for understanding ultrasonic backscattering from blood. *J. Acoust. Soc. Am.*, 98, 3040–3049.

[21] Bashkatova, T. A., Bashkatova, A. N., Kochubey, V. I., and Tuchin, V. V., 2001. Light scattering properties for spherical and cylindrical particles: A simple approximation derived from Mie calculations, *Proc. SPIE*, 4241, 247–258.

[22] Bassiri, S., 1987. Electromagnetic wave propagation and radiation in chiral media. Ph.D thesis, Caltech.

[23] Bates, R. H., James, J. R., Callet, I. N. L., and Miller, R. F., 1973. An overview of point matching. *Radio Elect. Eng.*, 43, 197–210.

[24] Bayvel, L. P., and Jones A. R., 1981. *Electromagnetic Scattering and Its Applications.* Appl. Sci., Essex.

[25] Bell, D. I., Noll, N. C., and Brown, E., 2003. CALC. C-style arbitrary precision calculator, version 2.22.10.1. http//www.isthe.com/chongo/tech/comp/calc/index.html.

[26] Ben-Zvi, S.Y., Connolly, B. M., Matthews, J. A. J., Prouza, M., Visbal, E. F., and Westerhoff, S., 2007. Measurement of the aerosol phase function at the Pierre Auger Observatory. *Astro. Part. Phys.*, 28, 312–320.

[27] Berg, M. J., 2012. Power-law patterns in electromagnetic scattering: A selected review and recent progress. *J. Quant. Spectrosc. Radiat. Transf.*, 113, 2292–2309.

[28] Berlad, G., 1971. An impact-parameter representation for all scattering angles. *Nuovo. Cim. A*, 6, 594–600.

[29] Berrocal, E., 2006. Multiple scattering of light in optical diagnostics of dense sprays and other complex turbid media. Ph.D. thesis, Cranefield University.

[30] Bevilacqua, F., and Depeursinge, C., 1999. Monte Carlo study of diffuse reflectance at source detector separations close to one transport mean free path length. *J. Opt. Soc. Am. A*, 16, 2935–2945.

[31] Bhandari, R., 1985. Scattering coefficients for a multi-layered sphere. *Appl. Opt.*, 24, 1960–1964.

[32] Bhandari, R., 1986. Tiny core or thin layer as a perturbation in scattering by a single layered sphere. *J. Opt. Soc. Am. A*, 3, 319–328.

[33] Bianchi, S., Ferrara, A., Giovanardi, C., 1996. Monte-Carlo simulation of dusty spiral-galaxies: Extinction and polarization properties. *Astrophys. J.*, 465, 127–144.

[34] Birchak, J. R., 1974. High dielectric constant microwave probes for sensing soil moisture. *Proc. IEEE*, 62, 93–98.

[35] Blank, A., 1955. *Trans. Chalmers Univ. Technol.* Gothenburg, 168.

[36] Blumer, H., 1925. Strahlungsdiagramme kleinder dielektrischer kugeln. *Zeitschrift für Physik*, 32, 119.

[37] Bodenschatz, N., Krauter, P., Liemert, A., and Kienle, A., 2016. Quantifying phase function influence in subdiffusively backscattering. *J. Biomed. Opt.*, 21, 035002.

[38] Bohren, C. F., 1974. Light scattering by an optically active sphere. *Chem. Phys. Lett.*, 29, 458–462.

[39] Bohren, C. F. and Gilra, D. P., 1979. Extinction by a spherical particle in an absorbing medium. *J. Colloid Interface Sci.*, 72, 215–221.

[40] Bohren, C. F., and Hunt, A. J., 1977. Scattering of electromagnetic waves by a charged sphere. *Can. J. Phys.*, 55, 1930–1935.

[41] Bohren, C. F. and Huffman, D. R., 1983. *Absorption and Scattering of Light by Small Particles.* Wiley, New York.

[42] Bohren, C. F., and Nevitt, T. J., 1983. Absorption by a sphere, a simple approximation. *Appl. Opt.*, 22, 774–775.

[43] Born, M., 1972. *Optik*. Berlin, Springer, p. 412.

[44] Borovoi, A. G., 2006. Multiple scattering of short waves by uncorrelated and correlated scatterers. *Light Scattering Reviews*, 1, 181–251.

[45] Bourrley, C., Chiapetta, P., and Lemaire, T., 1996. Improved version of the eikonal model for absorbing spherical particles. *J. Mod. Opt.*, 43, 409–415.

[46] Box, M. A., 1983. Power series expansion of the Mie scattering phase function. *Aust. J. Phys.*, 36, 710–716.

[47] Brillouin, L., 1926. La mechanique ondulatoire de Schroedinger: Une methode generale de resolution par approximations successives. *C. R. Acad. Sci.*, 183, 24–26.

[48] Brok, J. M., 2007. An analytic approach to electromagnetic scattering problems. Ph.D thesis, Univ. Delft.

[49] Brown, A. J., 2013. On the effects of size factor on albedo versus wavelength for light scattered by small particles under Mie and Rayleigh regimes. arXiv:1307.5096[physics.optics].

[50] Bruggeman, D. A. G., 1935. Berechnung vershiedener physikalischer konstanten von heterogenen substanzen. 1. Dilectrizitatskonstentan und leitfahigkeiten der mischkorper aus isotrpen substanzen. *Ann. Phys. (Leipzeg)*, 24, 636–664.

[51] Brunsting, A., and Mullaney, P. F., 1971. Light scattering from coated spheres: Model for biological cells. *Appl. Opt.*, 11,675–680.

[52] Bryant, H. C., and Cox, A. J., 1966. Mie theory and the glory. *J. Opt. Soc. Am.*, 56, 1529–1532.

[53] Bryant, F. D., and Latimer, P., 1969. Optical efficiencies of large particles of arbitrary shape and orientations. *J. Coll. Interf. Sci.*, 30, 291–304.

[54] Burberg, R., 1956. Die beugung electromagnetischer Wellen am unendlich laengen Kreiszylinder. *Z. Naturforsch*, 11A, 800–806.

[55] Cachorro, V. E. and Salcedo, L. L., 2001. New improvements for Mie scattering calculations. *J. Electromagnetic Waves and Applications*. 5, 913–926.

[56] Calabro, K. W., and Bigio, I. J., 2014. Influence of the phase function in generalized diffuse reflectance modes: Review of current formalisms and novel observations. *J. Biomed. Opt.*, 19, 075005.

[57] Caldas M and Semiã*o*, V., 2001a. Radiative properties of small particles. Extension of the Penndorf model. *J. Opt. Soc. Am. A*, 18, 831–838.

[58] Caldas M and Semião, V., 2001b. A new approximate phase function for isolated particles and polydispersions. *J. Quant. Spectrosc. Radiat. Transf.*, 68, 521–542.

[59] Cantrell, C. D., 1988 (references updated 2006). Numerical methods for accurate calculation of spherical Bessel functions and the location of Mie resonances. *Tech. Report, Univ. Texas.*

[60] Car, R., Cicuta, G. M., Zanon, D., and Riva, F., 1977. High energy, Rytov, eikonal expansions. *Nuovo Cim.*, 39, 253–271.

[61] Cardelli, J. A., Clayton, G. C., and Mathis, J. S., 1989. The relationship between infrared, optical and ultraviolet extinction. *Astrophy. J.*, 345, 245–256.

[62] Carlini, F., 1817. *Ricerche sulla convergenza della series che serva alla soluzionedel peoblema di Keplero.* Milano.

[63] Casperson L. W., 1977. Light extinction in polydisperse particulate systems. *Appl. Opt.*, 16, 3183–3189.

[64] Chamot, S., Migacheva, E., Seydoux, O., Marquet, P., and Depeursinge, C., 2010. Physical interpretation of the phase function related parameter γ studied with a fractal distribution of spherical scatterers. *Opt. Express*, 18, 23664–23675.

[65] Chandrasekhar, S., 1960. *Radiative Transfer.* Dover Publications, Inc., New York.

[66] Chen, T. W., 1987. Scattering of light by a stratified sphere. *Appl. Opt.*, 26, 4155–4158.

[67] Chen, T. W., 1989. High energy light scattering in the generalized eikonal approximation. *Appl. Opt.*, 28, 4096–4102.

[68] Chen, T. W., 1993. Simple formula for light scattering by a large spherical dielectric. *Appl. Opt.*, 32, 7568–7571.

[69] Chen, T. W., 1994. Diffraction by a spherical dielectric at large size parameter. *Opt. Commun.*, 107, 189–192.

[70] Chen, T. W., 1995. Effective sphere for spheroid in light scattering. *Opt. Commun.*, 114, 199–202.

[71] Chen, T. W., and Yang, L. M., 1996. Simple formula for small angle light scattering by a spheroid. *Opt. Commun.*, 123, 437–442.

[72] Chen, Z., Taflove, A., and Backman, V., 2003. Equivalent volume averaged light scattering behaviour of randomly inhomogeneous dielectric spheres in the resonant range. *J. Opt. Lett.*, 28, 764–767.

[73] Chen, Z., Taflove, A., and Backman, V., 2004. Concept of the equiphase sphere for light scattering by nonspherical dielectric particles. *J. Opt. Soc. Am. A*, 21, 88–97.

[74] Chernyshev , A. V., Maltsev, V. P., Prots, V. I., and Doroshkin, A. A., 1995. Measurement of scattering properties of individual particles with a scanning flow cytometer. *Appl. Opt.*, 34, 6301–6305.

[75] Chicea, D., and Chicea, L. M., 2006. On light scattering anisotropy of biological fluids (urine) characterization. *Rom. J. Phys.*, 52, 383–388.

[76] Chu, C. M., and Churchill, S. W., 1955. Representation of the angular distribution of radiation scattered by a spherical particle. *J. Opt. Soc. Am.*, 45, 958–962.

[77] Chylek, P., 1973. Mie scattering into the backward hemisphere. *J. Opt. Soc. Am.*, 63,1467–1471.

[78] Chylek, P., 1976. Partial-wave resonances and the ripple structure in the Mie normalized extinction cross section. *J. Opt. Soc. Am.*, 66, 285–287.

[79] Chylek, P., 1977. Light scattering by small particles in an absorbing medium. *J. Opt. Soc. Am. A*, 67, 561–563.

[80] Chylek, P., 1990. Resonance structure of Mie scattering: Distance between resonances. *J. Opt. Soc. Am. A*, 7, 1609–1613.

[81] Chylek, P., and Klett. J. D., 1991a. Extinction cross section of nonspherical particles in the anomalous diffraction approximation. *J. Opt. Soc. Am. A*, 8, 274–281.

[82] Chylek, P., and Klett. J. D., 1991b. Absorption and scattering of electromagnetic radiation by prismatic columns: Anomalous diffraction approximation. *J. Opt. Soc. Am. A*, 8, 1713–1720.

[83] Chylek, P., and Videen, G., 1994. Longwave radiative properties of polydisperse hexagonal ice crystals. *Atmos. Sci.*, 51, 175–190.

[84] Chylek, P., Grams, G. W., Smith, G. A., and Russel, P. B., 1975. Hemispherical backscattering by aerosols. *J. Appl. Met.*, 14, 380–387.

[85] Chylek, P., Kiehl, J. T., and Ko, M. K. W., 1978. Optical levitation and partial wave resonances. *Phys. Rev. A.*, 18, 2229–2233.

[86] Chylek, P., Videen, G., Wally Geldart, D. J., Steven Dobbie, J., and William Tso, H. C., 2000. Effective medium approximations for heterogeneous particles. In *Light scattering by nonspherical particles*. Eds. Mishchenko, M. I., Hovenier, J. W., and Travis, L. D., Academic Press, San Diego.

[87] Clark, G. C., Chu, C. M., and Churchill, S. W., 1957. Angular distribution coefficients for radiation scattered by a spherical particle. *J. Opt. Soc. Am.*, 47, 81–84.

[88] Coleman, R., 1981. Intercept lengths of random probes through boxes. *J. Appl. Prob.*, 18, 276–281.

[89] Condon, E. U., 1937. Theories of optical rotatory power. *Rev. Mod. Phys.*, 9, 432–57.

[90] Cooray, M. F. R., and Ciric, I. R., 1993. Wave scattering by a chiral spheroid. *J. Opt. Soc. Am. A.*, 10, 1197–1203.

[91] Cornette, W. M., and Shanks, J. G., 1992. Physically reasonable analytic expression for the single-scattering phase function. *Appl. Opt.*, 31, 3152–3160.

[92] Cross, D. A., and Latimer, P., 1970. General solutions for extinction and absorption efficiencies of arbitrarily oriented cylinders by anomalous diffraction approximation. *J. Opt. Soc. Am.*, 60, 904–907.

[93] Crossbie, A. L., and Davidson, G. W., 1985. Dirac-delta function approximations to the scattering phase functions. *J. Quant. Spectrosc. Radiat. Transf.*, 33, 391–409.

[94] Cubeddu, R., Pifferi, A., Taroni, P., Torricelli and di Milano, P., 2002. Measuring fresh fruit and vegetable quality: Advanced optical methods. In *Fruit and Vegetable Processing: Improving Quality.* Ed W. Jongen, CRC Press, Washington DC.

[95] Davis, E., and Schweiger, G., 2002. *The airborne microparticle.* Springer, Berlin.

[96] Davison, D., 1957. *Neutron Transport Theory.* Oxford Press, Clarendon.

[97] Debye, P., 1908, Das elektromagnetische field um einen zylinder und die theorie des regenbogens. *Phys. Zeit.*, 9, 775–778. Translated into English and reprinted in "Selected papers on geometrical aspects of scattering." Ed. by P. L. Martson, *SPIE Milestone Series*, MS89, 198 (1994).

[98] Debye, P., 1909. Der lichtdruck auf kugeln von beliebigem material. *Annalen der Physik*, Vierte Band 30. 57–136.

[99] Debye, P., 1915. Scattering from non-crystalline substances. *Ann. Physik.*, 46, 809–823.

[100] Denman, H. H., Heller, W., and Pangonis, W. J., 1966. *Angular scattering functions for spheres.* Wayne State University Press, Detroit.

[101] Deirmendjian, D., 1969. *Electromagnetic Scattering on Spherical Polydispersions.* Elsevier, New York.

[102] Devaux, C., Siewert, C. E., and Yener, Y., 1979. The effect of forward and backward scattering on the scattering albedo for a finite plane parallel atmosphere. *J. Quant. Spectrosc. Radiat. Transf.*, 21, 505–509.

[103] Di Marzio, F and Szajman, J., 1992. Mie scattering in the first order corrected eikonal approximation. *Computer Physics Commun.*, 70, 297–304.

[104] Dobbins, R. A., and Megaridis, C. M., 1991. Scattering and absorption of light by polydisperse aggregates. *Appl. Opt.*, 30, 4747–4754.

[105] Dombrovsky, L. A., 2002. A spectral model of absorption and scattering of thermal radiation by droplets of diesel fuel. *High Temp*, 40, 242–248.

[106] Dombrovsky, L. A., and Baillis, D., 2010. *Thermal Radiation in Disperse Systems: An Engineering approach.* Begell House, Inc., New York.

[107] Draine, B. T., 1988. The discrete-dipole approximation and its applications to interstellar graphite grains. *Astrophys. J.*, 333, 848–872.

[108] Draine, B. T., 2003. Scattering by interstellar dust grains. I. Optical and Ultraviolet. *ApJ.*, 598, 1017–1025.

[109] Draine, B. T., 2011. *Physics of the Interstellar and Intergalactic Medium.* Princeton University Press, Princeton.

[110] Draine, B. T., and Flatau P. J., 1984. Discrete dipole approximation for scattering calculations. *J. Opt. Soc. Am. A*, 11, 1491–1499.

[111] Draine, B. T., and Malhotra, S., 1993. On graphite and the 2175$\AA$ extinction profile. *Astrophys. J.*, 414, 632–645.

[112] Draine, B. T., and Allaf-Akbari, K., 2006. X-ray scattering by nonspherical grains: Oblate spheroids. *Astrophys. J.*, 652, 1318–1330.

[113] Drolen, B. L., and Tien, C. L., 1987. Absorption and scattering of agglomerated soot particulate. *J. Quant. Spectrosc. Radiat. Transf.*, 37, 433–448.

[114] Edwards, J. M., and Slingo, A., 1996. Studies with a flexible new radiation code . I. Choosing a configuration for large-scale model. *Q. J. R. Meteorol. Soc.*, 122, 689–719.

[115] Erdelyi, A. (Ed.), 1953. *Higher Transcendental Functions.* Vol. 1. McGraw Hill, New York, p 170.

[116] Evans, B. T. N., and Fournier, G. R., 1990. A simple approximation to extinction efficiency valid over all size parameters. *Appl. Opt.*, 29, 4666–4670.

[117] Evans, B. T. N., and Fournier, G. R., 1994. Analytic approximation to randomly oriented spheroid extinction. *Appl. Opt.*, 33, 5796–5804.

[118] Fang, S-h., Fu, X-p., and He, X-m., 2016. Investigation of absorption and scattering characteristics of kiwifruit tissue using a single integrating sphere system. *J Zhejiang Univ-Science B*, 17, 484–492.

[119] Farfonov, V. G., Il'in, V. B., and Prokopjeva, M. S., 2001. Scattering of light by homogeneous and multilayered ellipsoids in quasistatic approximation. *Opt. Spectrosc.*, 92, 608–617.

[120] Farafonov, V. G.,and Il'in, V. B., 2006. Single light scattering: Computational methods. *Light Scattering Reviews*, 1, 125–177.

[121] Farone, W. A., Kerker, M., and Matijevic, E., 1963. *Electromagnetic Scattering* (M. Kerker Ed.) pp. 55–71, Pergamon Press, Oxford.

[122] Fenn, R. N, and Oser, H., 1965. Scattering properties of soot water concentric spheres for visible and infrared light. *Appl. Opt.*, 4, 1504–1509.

[123] Ferdinandov, E. S., 1967. Light diffraction from a spherical particle of thin concentric shells. *Bulg. J. Phys.*, 3, 323–333.

[124] Fernandez-Oliveras, A., Rubino, M., and Perez, M. M., 2012. Scattering anisotropy measurements in dental tissues and biomaterials. *J. Euro. Opt. Soc. Rapid Publications*, 7, 12016-1-8.

[125] Fishburne, E. S., Neer, M. E., and Sandri, G., 1976. Voice communication via ultrasound radiation. Report 274. Vol. 1, Princeton, NJ. Aeronautical Research Associates of Princeton.

[126] Fitzpatrick, E. L., and Masa, D., 1988. An analysis of shapes of interstellar extinction curves. II-The far-UV extinction. *Astrophys. J.*, 328, 734–746.

[127] Fitzpatrick, E. L., and Masa, D., 1990. An analysis of the shapes of ultraviolet extinction curves. III. An atlas of ultraviolet extinction curves. *Astrophys. J. Supplement series*, 72, 163–189.

[128] Flatau, P. J., 1992. Scattering by irregular particles in anomalous diffraction and discrete dipole approximations. Atmospheric Science Paper No. 517. Colorado State University, Fort Collins.

[129] Flock, S. T., Wilson, B. C., and Patterson, M. S., 1987. Total attenuation coefficient and scattering phase function of tissues and phantom materials at 633 *nm*. *Med. Phys.*, 14, 826–834.

[130] Forand,J. L., and Fournier, G. R., 1999. Particle distribution and index of refraction estimation for Canadian waters. *Proc. SPIE*, 3761, 34–44.

[131] Fournier,, G. R., and Evans, B. T. N., 1991. Approximations to extinction from randomly oriented spheroids. *Appl. Opt.*, 30, 2041–2048.

[132] Fournier, G. R., and Evans, B. T. N., 1996. Approximations to extinction from randomly oriented circular and elliptic cylinders. *Appl. Opt.*, 35, 4271–4282.

[133] Fournier, G. R., and Forand, J. L., 1994. Analytic phase functions for Ocean waters. In *Proc. SPIE*, 2258, 194–201.

[134] Fowler, B. N., 1983. Expansion of Mie theory phase functions in series of Legendre polynomials. *J. Opt. Soc. Am.*, 73, 19–22.

[135] Fried, D., Glena, R. D., Featherstine, J. D. B., and Seka, W., 1995. Nature of light scattering in dental enamel and dentin at visible and near infrared wavelengths. *Appl. Opt.*, 7, 1278–1285.

[136] Frisvad, J. R, Christensen, N. J. and Jensen, H. W., 2007. Computing the scattering properties of participating media using Lorenz–Mie theory. *ACM Transactions on Graphics*, 26 vol, 3, article 60.

[137] Fu, Q., and Sun, W., 2001. Mie theory for light scattering by a spherical particle in an absorbing medium. *Appl. Opt.*, 40, 1354–1361.

[138] Fu, Q., and Sun, W., 2006. Apparent optical properties of spherical particles in absorbing medium. *J. Quant. Spectrosc. Radiat. Transf.*, 100, 137–142.

[139] Fu, Q., Sun, W. B., and Yang, P., 1999. Modelling of scattering and absorption by nonspherical cirrus ice particles at thermal infrared wavelengths. *J. Atmos. Sci.*, 56, 2937–2947.

[140] Fu, Q., Thorsen, T. J., Su, J., Ge, J. M., and Huang, J. P., 2009. Test of Mie based single-scattering properties of nonspherical dust aerosols in radiative flux calculations. *J. Quant. Spectrosc. Radiat. Transf.*, 110, 1640–1653.

[141] Fymat, A. L., and Mease, K. D., 1981. Mie forward scattering: Improved semiempirical approximation with application to particle size distribution inversion. *Appl. Opt.*, 20, 194–198.

[142] Gans, R., 1915. Fortplantzung des lichts durch ein inhomogenes medium. *Ann. Phys.* 352, 709–736.

[143] Gans, R., 1925. Strahlunsdiagramme ultramickroscopischer teilchen. *Annalen der Physik*, 381 29–38.

[144] Garcia-Camara, B., Gonzalez, F., Moreno, F., and Videen, G., 2008. Light scattering resonances in small particles with electric and magnetic optical properties. *J. Opt. Soc. Am. A*, 25, 327–334.

[145] Garcia-Camara, B., Saiz, J. M., Gonzalez, F., and Moreno, F., 2010. Nanoparticles with unconventional properties: Size effects. *Opt. Commun.*, 283, 490–496.

[146] Garcia-Camara, B., Alcaraz de la osa, R., Saiz, J. M., Gonzalez, F., and Moreno, F., 2011, Directionality in scattering by nanoparticles: Kerker's null scattering conditions revisited. *Opt. Lett.*, 36, 728–730.

[147] Garcia-Lopez, A. C., Snider, A.D. and Garcia-Rubio, L. H., 2006. Rayleigh–Debye–Gans as a model for continuous monitoring of biological particles: Part I, assessment of theoretical limits and approximations. *Opt. Exp.*, 14, 8850–8865.

[148] Garcia-Lopez, A. C.,and Garcia-Rubio, L. H., 2008. Rayleigh–Debye–Gans as a model for continuous monitoring of biological particles: Part II, development of a hybrid model. *Opt. Exp.*, 16, 4671–4687.

[149] Gelebart, B., Tinet, E, Tualle, J. M., and Avrillier, S., 1996. Phase function simulation in tissue phantoms: A fractal approach. *Pure Appl. Opt. A*, 5, 377–388.

[150] Geng, Y., and He, S., 2006. Analytical solution for electromagnetic scattering from a sphere of uniaxial left-handed material. *J. Zhejiang Univ.*, 7, 99–104.

[151] Geng, Y, Qiu, C., and Yuan, N., 2009. Exact solution of electromagnetic scattering by an impedance sphere coated with a uniaxial anisotropic layer. *IEEE Trans. AP*, 57, 572–576.

[152] Gille, W., 2000. Chord length distributions and small angle scattering. *Eur. Phys. J. B*, 17, 371–383.

[153] Glantsching, W.J., and S-H. Chen., 1981. Light scattering from water droplets in geometrical optics approximation. *Appl. Opt.*, 20, 2499–2509.

[154] Gobel, G., Kuhn, J., and Fricke, J., 1995. Dependent scattering effects in latex-sphere suspensions and scattering powders. *Waves in Random Media 5*, 4, 413–426.

[155] Gordon, D. J., 1972. Mie scattering by optically active particles. *Biochemistry*, 11, 413–420.

[156] Gordon, J. E., 1985. Simple method for approximating Mie theory. *J. Opt. Soc. Am.*, A2, 156–159.

[157] Gouesbet, G., 2003. Debye series formulation for generalized Lorenz–Mie theory with the Bromwich method. *Part. Part. Syst. Charact.*, 20, 382–386.

[158] Gouesbet, G., and Grehan, G., 1982. A generalized Lorenz–Mie theory. *J. Opt.*, 13, 97–103.

[159] Gouesbet, G., and Grehan, G., 2011. *Generalized Lorenz–Mie Theories.* Springer, Berlin.

[160] Grandy, W. T,. 2000. *Scattering of Waves from Large Spheres.* Cambridge University Press, Cambridge.

[161] Granovskii, Ya., I., and Ston, M., 1994a. Light scattering cross-sections: Summing of Mie series. *Physica Scripta*, 50, 140–141.

[162] Granovskii, Ya., I., and Ston, M., 1994b. Attenuation of light scattered by transparent particles. *JETP (USA)*, 78,645–649.

[163] Green, G., 1837. On the motion of a wave in a variable canal of small depth and width. *Trans. Cambridge Philos. Soc.*, 6, 457–462.

[164] Grenfell, T. C., and Warren, S. G., 1999. Representation of a nonspherical ice particle by a collection of independent spheres for scattering and absorption of radiation. *J. Geophys. Res. Atmos.* , 31, 697–709.

[165] Grenfell, T. C., Neshyba, S. P., and Warren, S. G., 2005 . Representation of a nonspherical ice particle by a collection of independent spheres for scattering and absorption of radiation. 3. Hollow columns and plates. *J. Geophys. Res. Atmos.*, 110:D17203.

[166] Gruy, F., 2014. Fast calculation of light differential scattering cross section of optically soft and convex bodies. *Opt. Commun.*, 313, 394–400.

[167] Guenther, B. D., 1990. *Modern Optics*, John Wiley and Sons, New York.

[168] Güttler, A., 1952. Die Mie theorie der Beugung durch dielektrische Kugeln mit absorbierendem kern und ihre Bedeutung für Probleme der interstellaren Materie und des atmospharischen Aerosols. *Ann. der Physik*, 11, 65–98.

[169] Haltrin, V. I., 1988. Exact solution of the characteristic equation for transfer in the anisotropically scattering and absorbing medium. *Appl. Opt.*, 27, 599–602.

[170] Haltrin, V. I., 2002. One-parameter two-term Henyey–Greenstein phase function for light scattering in seawater. *Appl. Opt.*, 41, 1022–1028.

[171] Hammer, M., Schweitzer, D., Micher, B., Thamm, E., and Kolb, A., 1998. Single scattering by red blood cells. *Appl. Opt.*, 37, 7410–7418.

[172] Hammer, M., Yaroslavsky, A. N., and Schweitzer, D., 2001. A scattering phase function for blood with physiological haematocrit. *Phys. Med. Biol.*, 46, N65–N69.

[173] Hansen, J. E., 1969. Exact and approximate solutions for multiple scattering by cloudy and hazy planetary atmospheres. *J. Atmos. Sci.*, 26, 478–487.

[174] Hapke, B. W., 1981. Bidirectional reflectance spectroscopy, I. Theory. *J. Geophys. Res.*, 86, 3039–3054

[175] Hart, R. W., and Montroll, E. W., 1951. On the scattering of plane waves by soft particles, I: Spherical particles. *J. Appl. Phys.*, 22, 376–386.

[176] Hartel, W., 1940. Zur theorie der lichtstruung durch trube schichten, besonders trubglaser. *Licht*, 40, 141–143.

[177] Heller, W., 1963. Theoretical and experimental investigation of light scattering colloidal spheres. In *Electromagnetic scattering.* Ed. M. Kerker, 107–120. Pergamon Press, Oxford.

[178] Heffels, C., Heitzmann, D., Hirleman, E. D., and Scarlett, B., 1995. Forward light scattering for arbitrarily sharp-edged crystals in Fraunhofer and anomalous diffraction approximation. *Appl. Opt.*, 34, 6552–6560.

[179] Henyey, L, and Greenstein, J., 1941. Diffuse radiation in the galaxy. *Astrophys. J.*, 93, 70–83.

[180] Hergert, W., and Wriedt, T. (Eds.), 2012. *The Mie Theory-Basics and Applications.* Springer, Heidelberg.

[181] Hinders, M., and Rhodes, B., 1992. Electromagnetic wave scattering from chiral spheres in chiral media. *Nuovo Cimento D*, 14, 575–583.

[182] Hoffman, J. and Draine, B. T., 2016. Accurate modelling of X-ray extinction by interstellar grains. *Astrophys. J.*, 817: 139: 1–15.

[183] Hong, N. S., 1977. A method of particle sizing using crossed laser beams. Ph.D thesis, Univ. London.

[184] Hovarth, H., 2009. Gustav Mie and the scattering and absorption of light by particles: Historic developments and basics. *J. Quant. Spectrosc. Radiat. Transf.*, 110, 787–799.

[185] Hu, Q., and Xie, L., 2015. Scattering phase function of a charged spherical particle. *Appl. Opt.*, 54, 8439–8443.

[186] Irvine, W. M., 1963. The asymmetry parameter of the scattering diagram of a spherical particle. *Bull. Astron. Inst.*, Netherlands, 3, 176–184.

[187] Ishimaru, A., 1999. *Wave Propagation and Scattering in Random Media.* Wiley-IEEE Press, New York.

[188] Ivanov, A.P., Loiko, V. A., and Dick, V. P., 1988. *Propagation of Light in Densely Packed Dispersive Media.* Nauka i Tekhnika, Minsk.

[189] Jacquier, S., and Gruy, F., 2008a. Anomalous diffraction approximation for light scattering cross section: Case of ordered clusters of non-absorbent spheres. *J. Quant. Spectrosc. Radiat. Transf.*, 109, 789–810.

[190] Jacquier, S., and Gruy, F., 2008b. Anomalous diffraction approximation for light scattering cross section: Case of random clusters of non-absorbent spheres. *J. Quant. Spectrosc. Radiat. Transf.*, 109, 2794–2803.

[191] Jacquier, S., and Gruy, F., 2010. Application of scattering theories to the characterization of precipitation processes. *Light Scattering Reviews*, 5, 37–78.

[192] Jacques, S. L., Alter, C. A., and S. A. Prahl, 1987. Angular dependence of He Ne laser light scattering by human dermis. *Laser Life Sci.,* 1, 309–333.

[193] Jaggard, D. L., Mickelson, A. R., and Papas, C. J., 1979. On electromagnetic waves in chiral media. *Appl. Phys.*, 18, 211–216.

[194] Jaggard, D. L., and Liu, J. C., 1999. The matrix Riccati equation for scattering from stratified chiral spheres. *IEEE Trans. Antennas Propag.*, 47, 1201–1207.

[195] Jerlov, N. G., 1968. *Optical Oceanography.* Elsevier, Amsterdam.

[196] Jobst, G., 1925. Diffuse strahlung dielektrischer kugeln im grenzfalle dass kugelmaterial und umgebendes medium fast gleiche brechungsindices haben. *Ann. Physik*, 78, 157–166.

[197] Johnson, B. R., 1993. Theory of morphology dependent resonances: Shape resonances and width formulas. *J. Opt. Soc. Am. A*, 10, 342–352.

[198] Jonasz, M., and Fournier, G. R., 2007. *Light Scattering by Particles in Water: Theoretical and Experimental Foundations.* Academic Press, New York.

[199] Jones, A. R., 1977. Error contour charts relevant to particle sizing by forward-scattered lobe methods. *J. Phys. D.*, 10, L163–L165.

[200] Jones, A. R., 1999. Light scattering for particle characterization. *Progr. Energ. Combust. Sci.*, 25, 1–53.

[201] Jones. A. R., Koh, J., and Nassarudin, A., 1996. Error contour charts for the two wave WKB approximation. *J. Phys. D* , 29, 39–42.

[202] Jones, D. S., 1957. High-frequency scattering of electromagnetic waves. *Proc. Roy. Soc. London*, A240, 206–213.

[203] Joseph, J. H., Wiscombe, W. J., and Wienman, J. S., 1976. The delta-Eddington approximation for radiative flux transfer. *J. Atmos. Sci.*, 33, 3452–2459.

[204] Kai, L., and Massoli, P., 1994. Scattering of electromagnetic plane waves by a radially inhomogeneous spheres: A finely stratified sphere model. *Appl. Opt.*, 33, 501–511.

[205] Kai, L., and D'Alessio, A., 1995. Finely stratified cylinder model for radially inhomogeneous cylinders normally irradiated by electromagnetic plane waves. *Appl. Opt.,* 34, 5520–5530.

[206] Kamiuto, K., 1987. Study of Henyey–Greenstein approximation to scattering phase function. *J. Quant. Spectrosc. Radiat. Transf.*, 37, 411–413.

[207] Kattawar, G. W., and Plass, G. N., 1967. Electromagnetic scattering from absorbing spheres. *J. Opt. Soc. Am.*, 6, 1377–1382.

[208] Kellerer, A. M., 1984. Chord-length distributions and related quantities for spheroids. *Radiat. Res.*, 98, 425–437.

[209] Kerker, M., 1969. *The Scattering of Light and Other Electromagnetic Radiation*. Academic Press. New York.

[210] Kerker, M., and Matijevic, E., 1961. Scattering of electromagnetic waves from concentric infinite cylinders. *J. Opt. Soc. Am.*, 51, 506–509.

[211] Kerker, M., Farone, W. A., and Matijevic, E., 1963. Applicability of the Rayleigh-Gans scattering to spherical particles. *J. Opt. Soc. Am.*, 53, 758–759.

[212] Kerker, M., Wang, D.-S., and Giles, C. L., 1983. Electromagnetic scattering by magnetic spheres. *J. Opt. Soc. Am.*, A73, 765–767.

[213] Kienle, A., Forester, A. K., and Hibst, R., 2001. Influence of phase function on determination of the optical properties of biological tissues by spatially resolved reflectance. *Opt. Lett.*, 26, 1571–1573.

[214] Kim, J. S., and Chang, J. K., 2004. Light scattering by two concentric optically active spheres. 1. General theory. *J. Korean Phys. Soc.*, 45, 352–365.

[215] Kim, C., and Lior, N., 1995. Easily computable good approximations for spectral radiative properties of particle-gas components and mixtures in pulverized coal combustors. *Fuel* , 74, 1891–1902.

[216] Kim, C., Lior, N., and Okuyama, K., 1996. Simple mathematical expressions for spectral extinction and scattering properties of small size-parameter particles, including examples for soot and TiO_2. *J. Quant. Spectrosc. Rad. Transf.*, 55, 391–411.

[217] Kitchen, J. C., and Zaneveld, J. R. V., 1992. A three layered sphere model of the optical properties of phytoplankton. *Limnol. Oceanogr.*, 37, 1680–1690.

[218] Klacka, J., and Kocifaj, M., 2007. Scattering of electromagnetic waves by charged spheres and some physical consequences. *J. Quant. Spectrosc. Radiat. Transf.*, 106, 170–183.

[219] Klacka, J., and Kocifaj, M., 2010. On the scattering of electromagnetic waves by a charged sphere. *Progress in Electromagnetic Research*, 109, 17–35.

[220] Klett, J. D., 1984. Anomalous diffraction model for inversion of multispectral extinction data including absorption effects. *Appl. Opt.*, 23, 4499–4508.

[221] Klett, J. D., and Sutherland, R. A., 1992. Approximate methods for modeling the scattering properties of nonspherical particles: Evaluation of the Wentzel–Kramers–Brillouin method. *Appl. Opt.*, 31, 373–386.

[222] Kocifaj, M., 2011. Approximate analytical scattering phase function dependent on microphysical characteristics of dust particles. *Appl. Opt.*, 50, 2493–2499.

[223] Kocifaj, M., Klacka, J., and Videen, G., 2011. Electromagnetic scattering by a polydispersion of small charged cosmic dust particles. *Atti Accad. Pelorit. dei Pericol. Cl. Sci. Fis. Mat. Nat.*, 89 (suppl. No. 1), 1–5.

[224] Kocifaj, M., and Klacka, J., 2012. Scattering of electromagnetic waves by charged spheres: Near-field external intensity distribution. *Opt. Lett.*, 37, 265–267.

[225] Kokhanovsky, A., 2002. Analytic solutions to multiple light scattering problems: a review. *Measurement Science and Technology*, 13, 233–240.

[226] Kokhanovsky, A., 2004. *Light Scattering Media Optics: Problems and Solutions*. Springer-Praxis, Chichester.

[227] Kokhanovsky, A., 2006. *Cloud Optics*. Springer, Netherlands.

[228] Kokhanovsky, A., 2008. *Aerosol Optics*. Springer-Verlag, Berlin.

[229] Kokhanovsky, A., 2008a. Phase functions of mixed clouds. *Atmos. Res.*, 89, 218–221.

[230] Kokhanovsky, A. A., and Zege, E. P., 1995. Local optical parameters of spherical polydispersions. *Appl. Opt.*, 34, 5573–5579.

[231] Kokhanovsky, A. A., and Zege, E. P., 1997. Optical properties of aerosol particles: A review of approximate analytic solutions. *Aerosol Sci.*, 28, 1–21.

[232] Komar, L., Kocifaj, M., and Kohut, I., 2013. An extension to the Rayleigh–Gans formula: Model of partially absorbing particles. *Optica Applicata*, 43, 313–323.

[233] Kramers, H. A., 1926. Wellemmechanik und halbzahlige quantisierung. *Z. Phys.*, 38, 518–529.

[234] Krugel. E., 2007. *An Introduction to the Physics of Interstellar Dust.* CRC Press, Boca Raton.

[235] Ku, J. C., and Felske, J. D., 1984. The range of validity of Rayleigh limit for computing Mie scattering and extinction efficiencies. *J. Quant. Spectrosc. Radiat. Transfer*, 31, 569–574.

[236] Kuwata, H., Tamaru, H., Esumi, K., and Miyano, K., 2003. Resonant light scattering from metal nanoparticles: Practical analysis beyond Rayleigh approximation. *Appl. Phys. Lett.*, 83, 4625–4628.

[237] Lakhtakia, A., Vardan, V. K., and Vardan, V. V., 1985. Scattering and absorption characteristics of lossy dielectric, chiral, nonspherical objects. *Appl. Opt.*, 24, 4146–4154.

[238] Lakhtakia, A., and Vikram, C. S., 1993. Variations of effective refractive index of a particulate composite. *Opt. Engg.*, 8, 1996–1998.

[239] Lam, C. C., Leung, P. T., and Young, K., 1992. Explicit asymptotic formulas for the positions, widths, and strengths of resonances in Mie scattering. *J. Opt. Soc. Am.*, A 9, 1585–1592.

[240] Landau, L. D., and Lifshitz, E. M., 1960. *Electrodynamics of Continuous Media.* pp. 26–27, Pergamon Press, Oxford.

[241] Latimer, P., 1975. Light scattering by ellipsoids. *J. Coll. Interf. Sci.*, 53, 102–109.

[242] Latimer, P., and Barber, P., 1978. Scattering of ellipsoids of revolution: A comparison of theoretical methods. *J. Coll. Interf. Sci.*, 633, 310–316.

[243] Latimer, P., 1980. Predicted scattering by spheroids: Comparison of approximate and exact methods. *Appl. Opt.*, 19, 3039–3041.

[244] Lebedev, A. N., Gartz, M., Kreibig, U., Stenzel, O., 1999. Optical extinction by spherical particles in an absorbing medium: Application to composite absorbing films. *J. Eur. Phys. D*, 6, 365–373.

[245] Lenekar, J., 1996. Optical properties of isotropic chiral media. *Pure Appl. Opt.*, 5, 417–443.

[246] Levine, P. H., 1978. Absorption efficiencies for large spherical particles: A new approximation. *Appl. Opt.*, 17, 3861–3862.

[247] Lentz, W. J., 1976. Generating Bessel functions in Mie scattering calculations using continued fractions. *Appl. Opt.*, 15, 668–671.

[248] Liang, S., and Townshend, J. R. G., 1996. A modified Hapke model for soil bidirectional reflectance. *Remote Sens. Environ.*, 55, 1–10.

[249] Li, L., Dan, Y., Leong, M., and Kong, J., 1999. Electromagnetic scattering by an inhomogeneous chiral sphere of varying permittivity: A discrete analysis using multilayered model. *Prog. Electromagn. Res.*, 13, 1203–1206.

[250] Li, J. L-W., Li, Z-C., She, H-Y., Zouhdi, S., Mosig, J. R., and Martin, O. J. F., 2009. A new closed form analytical solution to light scattering by spherical nanoshells. *IEEE Trans. Nanotechnology*, 8, 617–612.

[251] Li, H., Liu, C., Bi, L., Yang, P., and Kattawar, G. W., 2010. Numerical accuracy of "equivalent" spherical approximations for computing ensemble-averaged scattering properties of fractal soot aggregates. *J. Quant. Spectrosc. Radiat. Transf.*, 111, 2127–2132.

[252] Li, X., Xie, L., and Zheng, X., 2012. The comparison between the Mie theory and the Rayleigh approximation to calculate the EM scattering by partially charged sand. *J. Quant. Spectrosc. Radiat. Transf.*, 113, 251–258.

[253] Li, N., Zhu, Y., Wang, Z., 2015. A discussion on the applicable condition of Rayleigh scattering. *IJRSA*, 5 62–66.

[254] Li, X., Min, X., and Liu, D., 2014. Rayleigh approximation for the scattering of small partially charged sand particles. *J. Opt. Soc. Am. A*, 31, 1495–1501.

[255] Lichtenecker, K., 1926. Die Dielektrizitätskonstante natürlicher und künstlicher Mischkörper. *Phys. Z*, 27, 115–158.

[256] Liouville, J., 1837. Sur le developpement des fonctions et series. *J. Math. Pures Appl.*, 1, 16–37.

[257] Lind, A. C., and Greenberg, J. M., 1966. Electromagnetic scattering by obliquely oriented spheroids. *J. Appl. Phys.*, 37, 3195–3203.

[258] Liu, P., 1994. A new phase function approximating to Mie scattering for radiative transport equation. *Phys. Med. Biol.*, 39, 1025–1036.

[259] Liu, C., Jonas, P. R., and Saunders, C. P. R., 1996. Accuracy of anomalous diffraction approximation to light scattering by column-like ice crystals. *Atmos. Res.*, 41, 63–69.

[260] Liu, Q., and Weng, F., 2006. Combined Henyey–Greenstein and Rayleigh phase function. *Appl. Opt.*,45, 7475–7479.

[261] Liu, Y., Arnott, W. P., and Hallet, J., 1998. Anomalous diffraction theory for arbitrarily oriented finite circular cylinders and comparison with exact T-matrix results. *Appl. Opt.*, 37, 5019–5029.

[262] Liu, W., Zou, D., and Xu, Z., 2015. Modelling of optical wireless scattering communication channels over broad spectra. *J. Opt. Soc. Am. A*, 32, 486–490.

[263] Lock, J. A., 1988. Cooperative effects among partial waves in Mie scattering. *J. Opt. Soc. Am. A*, 5, 2032–2044.

[264] Lock, J. A., 2005. Debye series analysis of scatting of a plane wave by a spherical Bragg grating. *Appl. Opt.*, 44, 5594–5603.

[265] Lock, J. A., and Adler C. L., 1997. Debye series analysis of the first-order rainbow produced in scattering of a diagonally incident plane wave by a circular cylinder. *J. Opt. Soc. Am. A*, 14, 1316–1328.

[266] Lock, J. A., Jamison, J. M., and Lin C.-Y., 1994. Rainbow scattering by a coated sphere. *Appl. Opt.*, 33, 4677–4690.

[267] Logan, N. A., 1965. Survey of some early studies of of the scattering of plane waves by a sphere. PROC. IEEE, 53, 773–785.

[268] Loiko, V. A., Krakhalev, M. N., Konkolovich, A. V., Prishchepa, O. O., Miskevich, A. A., and Zyryanov, V. Ya., 2016. Experimental results and theoretical model to describe angular dependence of light scattering by monolayer of nematic droplets. *J. Quant. Spectrosc. Radiat. Transf.*, 178, 263–268.

[269] Looyenga, H., 1965. Dielectric constants of heterogeneous mixtures. *Physica*, 31, 401–486.

[270] Lopatin, V. N., and Sid'ko, F. Ya., 1988. *Introduction to the Optics of Cell Suspensions.* Nauka, Novosibirsk.

[271] Lorentz, H. A., 1880. Ueber die beziehung zwischen der fortpflanzungsgeschwindigkeit des lichtes und der korperdichte. *Ann. Phys. Chem.*, 9, 641–665.

[272] Lorenz, L. V., 1880. Ueber die refractionconstante. *Ann. Phys. Chem.*, 11,70–103.

[273] Lorenz, L. V., 1890. On the light refracted and refracted by a transparent sphere. *Vidensk. Selsk. Skr.*, 6, 1–62.

[274] Louedec, K., and Urban, M., 2012. Ramsauer approach for light scattering on non-absorbing spherical particles and applications to Henyey–Greenstein phase function. *Appl. Opt.*, 51, 7842–7852.

[275] Love, A. E. N., 1899. Scattering of electric waves by a dielectric sphere. *Proc. Lond. Math. Soc.*, 30, 308–321.

[276] Lowan, A. N., 1949. *Tables of Scattering Functions for Spherical Particles. Natl. Bu. Stds. Appl. Math.*, Ser. 4, Washinton DC.

[277] Lu, R. (Ed.), 2016. *Light Scattering Technology for Food Property, Quality and Safety Assessment.* CRC Press, Boca Raton.

[278] Ludlow, I. K., and Everitt, J., 1995. Application of Gegenbauer analysis to light scattering from spheres: Theory. *Phys. Rev E*, 51, 2516–2526.

[279] Ludlow, I. K., and Everitt, J., 1996. Systematic behaviour of Mie scattering coefficients of spheres as a function of order. *Phys. Rev. E*, 53, 2909–2924.

[280] Ludlow, I. K., and Everitt, J., 2000. Inverse Mie problem. *J. Opt. Soc. Am. A*, 17, 2229–2235.

[281] Lumme, K., and Rahola, J., 1994. Light scattering by porous dust particles in the discrete-dipole approximation. *Astrophys. J.*, 425, 653–667.

[282] Mäder, U., 1980. Chord length distributions for circular cylinders. *Radiat. Res.*, 82, 454–466.

[283] Mahood, R. W., 1987. The application of the vector diffraction to the scalar anomalous diffraction approximation of van de Hulst. Masters thesis, Pennsylvania State University, Dept. of Meteorology, PA.

[284] Malinka, A. V., 2015. Analytical expressions for characteristics of light scattering by arbitrary shaped particles in the WKB approximation. *J. Opt. Soc. Am. A*, 32, 1344–1351.

[285] Maltsev, V. P., 2000. Scanning flow cytometry for individual particles. *Rev. Sci. Inst.*, 71, 243–255.

[286] Maltsev, V. P., and Lopatin, V. N., 1997. Parametric solutions to the inverse light-scattering problem for individual spherical particles. *Appl. Opt.*, 36, 6102–6108.

[287] Marchesini, R., Bertoni, A., Andreola, S., Melloni, E., and Scichirollo, A. E., 1989. Extinction and absorption coefficients and scattering phase functions of human tissues *in Vitro. Appl. Opt.*, 28, 2318-2324.

[288] Marks, D. L., 2006. A family of approximations spanning the Born and Rytov scattering series. *Opt. Express*, 14, 8837–8848.

[289] Maslowska, A., 1991. Interaction of light with particles. *Acta. Geophys. Polonica*, 39, 113–128.

[290] Maslowska, A., Flatau, P. J., and Stephens, G. L., 1994. On the validity of anomalous diffraction theoryto light scattering by cubes. *Opt. Commun.*, 107, 35–40.

[291] Matciak, M., 2012. Anomalous diffraction approximation to the light scattering coefficient spectra of marine particles with power-law size distribution. *Opt. Express*, 27603-27611.

[292] Mathis, J. S., Rumpl, W., and Nordsieck, K. H., 1977. The size distribution of interstellar grains. *Astrophys. J.*, 217, 425–433.

[293] Mauche, C. W., and Gorenstein, P., 1986. Measurements of X-ray scattering from interstellar grains. *Astrophys. J.*, 302, 371–387.

[294] Maxwell–Garnett, J. C., 1904. Colours in metal glasses and in metallic films. *Philos. Trans. R. Soc. A*, 203, 385–420.

[295] McCormick, N. J., 1987. Inverse radiative transfer with a delta-Eddington phase function. *Astrophys. Space Sci.*, 129,331–334.

[296] McCormick, N. J., and Rinaldi, G. E., 1989. Seawater optical property estimation from in situ measurements. *Appl. Opt.*, 28, 2605–2613.

[297] Mehta, R. V., Patel, R., Desai, R., Upadhyay, R. V., and Parekh, K., 2006. Experimental evidence of zero forward scattering by magnetic spheres. *Phys. Rev. Lett.*, 96, 127402-1-127402-4.

[298] Meier, M., and Wokaun, A., 1983. Enhanced fields on large metal particles: Dynamic depolarization. *Opt. Lett.*, 11, 581–583.

[299] Mie, G., 1908. Beiträge zur optik truber medien speziell kolloidaler metallösungen. *Ann Physik*, 25, 377–445.

[300] Mikulski, J. J., and Murphy, E. L., 1963. The computation of electromagnetic scattering from concentric spherical structures, *IEEE Trans. Antenn. Propagat.*, AP-11, 169–177.

[301] Min, M., Hovenier, J. W., Diminik, C., de Koter, A., and Yurkin, M. A., 2006. Absorption and scattering properties of arbitrarily shaped particles in the Rayleigh domain: A rapid computational method and a theoretical foundation for statistical approach. *J. Quant. Spectrosc. Radiat. Transf.*, 97, 161–180.

[302] Miroshnichenko, A. E., 2009. Non-Rayleigh limit of the Lorenz–Mie theory solution and suppression of scattering by spheres of negative refractive index. *Phys. Rev. A*, 80, 013808.

[303] Mishchenko, M. I., 1994. Asymmetry parameters of the phase functions for densely packed scattering grains. *J. Quant. Spectrosc. Radiat. Transf.*, 52, 95–110.

[304] Mishchenko, M. I., 2009. Gustav Mie and the fundamental concepts of electromagnetic scattering by particles: A perspective. *J. Quant. Spectrosc. Radiat. Transf.*, 110, 1210–1222.

[305] Mishchenko, M. I., and Macke, A., 1997. Asymmetry parameters of the phase function for isolated and densely packed spherical particles with multiple internal inclusions in the geometric optics limit. *J. Quant. Spectrosc. Radiat. Transf.*, 57, 767–794.

[306] Mishchenko, M. I., Travis, L. D., and Mackowski, D. W., 1996. T-matrix computations of light scattering by nonspherical particles: A review. *J. Quant. Spectrosc. Radiat. Transf.*, 55, 535–575.

[307] Mishchenko, M. I., Dlugach, J. M., Yanovitskij, E. G., and Zakharova, N. D., 1999. Bidirectional reflectance of flat, optically thick particle layers: An efficient radiative transfer solution and applications to snow and soil surfaces. *J. Quant. Spectrosc. Radiat. Transf.*, 63, 409–432.

[308] Mishchenko, M. I., Travis, L. D. and Lacis, A. A., 2002. *Scattering, absorption, and emission of light by small particles.* Cambridge University Press, Cambridge.

[309] Mishchenko, M. I., Travis, L. D. and Lacis, A. A., 2006. *Multiple scattering of light by particles: Radiative transfer and coherent backscattering.* Cambridge University Press, Cambridge.

[310] Mishchenko, M. I., Liu, L., and Videen, G., 2007. Conditions of the applicability of single scattering approximation. *Optics Express*, 15, 7522–7527.

[311] Mobley, D., Sundman, L. K., and Boss, E., 2002. Phase function effects on oceanic light fields. *Appl. Opt.*, 41, 1135–1150.

[312] Moeglich, F., 1927. Beugungserscheinungen an korpen von ellipsoidischer gestalt. *Ann. Physik*, 83, 609–735.

[313] Montroll, E W and Hart R. W., 1951. On the scattering of waves by soft particles, II. Scattering by spheroids, cylinders and disks. *J. Appl. Phys.*, 22, 1278–1289.

[314] Monzon, J. C., 1989. Three-dimensional field expansion in most general rotationally symmetric anisotropic material: Scattering by a sphere. *IEEE Trans. Anten. Propag.*, 37, 728–735.

[315] Morris, V. J., and Jennings, B. R., 1977. Anomalous diffraction approximation to the low-angle light scattering from coated spheres. *Biophys. J.*, 17, 95–101.

[316] Morse, P. M., and Ingard, K. U., 1968. *Theoretical Acoustics.* McGraw Hill, New York.

[317] Mundy, W. C., Roux, J. A. and Smith, A. M., 1974. Mie scattering by spheres in an absorbing medium. *J. Opt. Soc. Am.*, 64, 1593–1597.

[318] Neves A. A. R. and Pisignano, D., 2012. Effect of Mie terms on truncation error of Mie series. *Opt. Lett.*, 37, 2418–2420.

[319] Newton, R. G., 1966. *Scattering Theory of Waves and Particles.* McGraw Hill, New York.

[320] Ngo, D., Videen, G., and Chylek, P., 1996. A FORTRAN code for scattering of EM waves by a sphere with a nonconcentric spherical inclusion. *Comput. Phys. Commun.*, 99, 94–112.

[321] Nicholls, R. W., 1984. Wavelength dependent spectral extinction of atmospheric aerosols. *Appl. Opt.*, 23, 1142–1143.

[322] Niemz, M. H., 2003. *Laser-Tissue Interactions: Fundamentals and Applications.* Springer-Verlag, Berlin.

[323] Nikolai, B. M., et al. 2014. Nondestructive measurement of fruit and vegetable quality, *Annual Reviews of Food Science and Technology*, 5: 285–312.

[324] Niklasson, G. A., and Granqvist, C. G., 1984. Optical properties of solar selectivity of coevaporated $Co - Al_2O_3$ composite films. *J. Appl. Phys.*, 55, 3382–3410.

[325] Nussenzweig, H. M., 1969a. High frequency scattering by a transparent sphere. I Direct reflection and transmission. *J. Math. Phys.*, 10, 82–124.

[326] Nussenzweig, H. M., 1969b. High frequency scattering by a transparent sphere. II Theory of the rainbow and glory. *J. Math. Phys.*, 10, 125–176.

[327] Nussenzweig, H. M., 1992. *Diffraction Effects in Semiclassical Scattering.* Cambridge University Press, Cambridge.

[328] Nussenzweig, H. M., and Wiscombe, W. J., 1980. Efficiency factors in Mie scattering. *Phys. Rev. Lett.*, 45, 1490–1494.

[329] Paramonov, L. E., 1994. On the optical equivalence of randomly oriented ellipsoidal and polydisperse spherical particles: The extinction, scattering and absorption cross sections. *Opt. Spectrosc.*, 77, 660–663.

[330] Paramonov, L. E., 2012. Optical equivalence of isotropic ensemble of ellipsoidal particles in the Rayleigh–Gans–Debye and anomalous diffraction approximations and its consequences. *Opt. Spectrosc.*, 112, 787–795.

[331] Paramonov, L. E., Lopatin V. N., Sidko, F. Ya., 1986. Light scattering by soft spheroidal particles. *Opt. Spectrosc.*, 61, 570–576.

[332] Passos, D., Hebden, J. C., Pinto, P. N., Guerra, R., 2005. Tissue phantom for optical diagnostics based on suspension of microspheres with a fractal dimension. *J. Biomed. Opt.*, 640361.

[333] Pegoraro, V, Schott, M, and Parker, S. G., 2010. A closed form solution to single scattering for general phase functions and light distributions. *Eurographics Symposium on Rendering*, 29, 1365–1374.

[334] Penndorf, R. B., 1958. An approximation method to the Mie theory for colloidal spheres. *J. Phys. Chem.*, 62, 1537–1542.

[335] Penndorf, R. B., 1962. Scattering and absorption coefficients for small absorbing and nonabsorbing aerosols. *J. Opt. Soc. Am.*, 52, 896–904.

[336] Perelman, A. Y., 1978. An application of Mie series to soft particles. *Pure Appl. Geophys.*, 116, 1077–1088.

[337] Perelman, A. Y., 1991. Extinction and scattering by soft particles. *Appl. Opt.*, 30, 475–484.

[338] Perelman, A. Y., 1994. Improvement of the convergence of a series for soft particle absorption cross section. *Opt. Spectrosc.*, 77, 643–647.

[339] Perelman, A. Y., 1996. Scattering by particles with radially variable refractive indices. *Appl. Opt.*, 35, 5452–5460.

[340] Perelman, A. Y., and Voshchinnikov, N. V., 2002. Improved S-approximation for dielectric particles. *J. Quant. Spectrosc. Radiat. Transf.*, 72, 607–621.

[341] Perrin, J. M., and Chiapetta, P., 1985. Light scattering by large particles, I. A new theoretical description of the eikonal picture. *Opt. Acta*, 32, 907–921.

[342] Petrov, D., Shakuratov, Y., and Videen, G., 2007. Analytical light-scattering solution for Chebyshev particles. *J. Opt. Soc. Am. A*, 24, 1103–1119.

[343] Petzold, J., 1972. Volume scattering functions for selected ocean waters. *Tech. Rep.*, SIO 72-78, Scripps Institution for Oceanography, San Diego, CA.

[344] Pfeiffer, N., and Chapman, G. H., 2008. Successive order, multiple scattering of two term Henyey–Greenstein phase unction. *Opt. Express*, 16, 13637–13642.

[345] Pinchuk, V. P., Romanov, N. P., 1977. Absorption cross section for spherical particles of arbitrary size with moderate absorptance. *J. Appl. Spectr.*, 27, 109–116.

[346] Piskozub, J., and McKee, D., 2011. Effective scattering phase functions for the multiple scattering regime. *Opt. Express*, 19, 4786–4794.

[347] Pomraning, G. C., 1988. On the Henyey–Greenstein approximation to the scattering phase function. *J. Quant. Spectrosc. Radiat. Transf.*, 39, 109–113.

[348] Pontikis, C., 1968. Retro diffusion d'une onde electromagnetique sur des obstacles spherique composes de plusieurs spheres concentriques. *Meteirologie*, N7, 271-283.

[349] Posselt, B., Farfonov, V. G., Il'in, V. B., and prokopjeva, M. S., 2002, Light scattering by multilayered spheroidal particles in quasistatic approximation. *Meas. Sci. Technol.*, 13, 256–262.

[350] Potter, J. F., 1970. The delta function approximation in radiative transfer theory. *J. Atmos. Sci.*, 27, 943–949.

[351] Prishivalko, A. P., Astafyeva, L. G., and Gladkaya, S. V., 1975. Investigation of light scattering and extinction by two layered particles. *Izvestiya AN BSSR. Ser. Fiz-mat. Nauk*, N3, 88–95.

[352] Probert-Jones, J. R., 1984. Resonance component of backscattering by large dielectric spheres. *J. Opt. Soc. Am. A*, 1, 822–830.

[353] Purcell, E. M., and Pennypacker, C. R., 1973. Scattering and absorption of light by nonspherical dielectric grains. *Astrophys. J.*, 186, 705–714.

[354] Quinten, M., 2011.*Optical Properties of Nanoparticle Systems: Mie and Beyond.* Wiley-VCH, Weinheim.

[355] Quinten, M, and Rostalski, J., 1996. Lorenz–Mie theory for spheres immersed in an absorbing host medium. *Particle & Particle Systems*, 13, 89–96.

[356] Quirantes, A., and Bernard, S., 2004. Light scattering by Marine algae: Two layer spherical and nonspherical models. *J. Quant. Spectrosc. Radiat. Transf.*, 89, 311–321.

[357] Räisänen, P., Kokhanovsky, A., Guyot, G., Jourdan, O., and Nousiainen, T., 2015. Parameterization of single-scattering properties of snow. *The Cryosphere*, 9, 1277–1301.

[358] Ramsauer, C. W., 1921. *Ü*ber den Wirkungsquerschnitt der gasmolek*ü*le gegen*ü*ber langsamen elektronen. *Ann. Physik*, 64, 513–540.

[359] Randrianalisoa, J., Baillis, D., and Pilon, L., 2006. Modeling radiation characteristics of semitransparent media containing bubbles or particles. *J. opt. Soc. Am. A*, 23, 1645–1656.

[360] Rayleigh, J. W. S., 1871. On the scattering of light by small particles. *Phil. Mag.*, S4, 41, 447–454.

[361] Rayleigh, J. W. S., 1872. Investigation of the disturbance produced by a spherical obstacle on the waves of sound. *Proc. Math. Soc. (London)*, 4, 253–283.

[362] Rayleigh, J. W. S., 1881. On the electromagnetic theory of light. *Phil. Mag.*, S5, 12, 81–101.

[363] Rayleigh, J. W. S., 1912. On the propagation of waves through a stratified medium, with special reference to the question of reflection. *Proc. R.Soc. A*, 86, 207–226.

[364] Reading, J. F., and Bassichis W. H., 1972. High energy scattering at backward angles. *Phys. Rev. D*, 5, 2031–2041.

[365] Reynolds, L. O., McCormick, N. J., 1980. Approximate two parameter phase function for light scattering. *J. Opt. Soc. Am.*, 70, 1206–1212.

[366] Riewe, F., and Green, A. E. S., 1978. Ultraviolet aureole around source at a finite distance. *Appl. Opt.*, 17, 1923–1929.

[367] Rogers, J. D., Capoglu, I. R., and Backman, V., 2009. Nonscalar elastic light scattering from continuous random media in the Born approximation. *Opt. Lett.*, 34, 1891–1893.

[368] Roy, A. K., and Sharma, S. K., 1996. On the validity of soft particle approximations for the light scattering by a homogeneous dielectric sphere. *J. Mod. Opt.*, 43, 2225–2237.

[369] Roy, A. K., and Sharma, S. K., 2005. A simple analysis of the extinction spectrum of a size distribution of Mie particles. *J. Opt. A: Pure Appl Opt.*, 7, 675–684.

[370] Roy, A. K., and Sharma, S. K., 2008. A simple parametrized phase function for small nonabsorbing spheres. *J. Quant. Spectrosc. Radiat. Transf.*, 109, 2804–2812.

[371] Roy, A. K., Sharma, S. K., and Gupta, R., 2009. A study of frequency and size distribution dependence of extinction for astronomical silicate and graphite grains. *J. Quant. Spectrosc. Radiat. Transf.*, 110, 1733–1740.

[372] Roy, A. K., Sharma, S. K., and Gupta, R., 2010. Frequency and size dependence of visible and infrared extinction for astronomical silicate and graphite grains. *J. Quant. Spectrosc. Radiat. Transf.*, 111, 795–801.

[373] Ru, E. C. L., Somerville, R. C., Auguié, B., 2013. Radiative correction in appropriate treatments of electromagnetic scattering by point and body scatterers. *Phys. Rev. A*, 87, 012504(1–12).

[374] Rutherford, E., 1911. The scattering of α and β particles by matter and the structure of the atom. *Philos. Mag.*, 21, 669–688.

[375] Rysakov, V. M., 2004. Light scattering by "soft" particles of arbitrary shape and size. *J. Quant. Spectrosc. Radiat. Transf.*, 87, 261–287.

[376] Rysakov, V. M., 2006. Light scattering by "soft" particles of arbitrary shape and size. II. Arbitrary orientation of particles in the shape. *J. Quant. Spectrosc. Radiat. Transf.*, 98, 85–100.

[377] Rytov, S. M., 1937. Diffraction of light by ultrasonic waves. *Izv. Akad. Nauk SSSR Ser. Fiz.*, 2, 223–259.

[378] Sahu, S. K., and Shanmugam, P., 2015. Semi-analytic modelling and parameterization of particulates-in-water phase function for forward angles. *Opt. Express*, 23, 22291–22307.

[379] Saxon, D. S., 1955. Lectures on scattering of light. *Scientific Report*, No. 9. UCLA, Los Angeles.

[380] Saxon, D. S., and Schiff, L. I., 1957. Theory of high energy potential scattering. *Nuovo Cim.*, 6, 614–627.

[381] Schaudt, K. J., Kwong, N. H., and Garcia, J. D., 1991. Exact solutions of light scattering from dielectric-disk arrays. *Phys. Rev. A*, 44, 4076–4079.

[382] Schebarchov, D., Auguié, B, Ru, E. C. L., 2013. Simple accurate approximations for optical properties of metallic nanospheres and nanoshells. *Phys. Chem. Chem. Phys.*, 15, 4233–4242.

[383] Schiff, L. I., 1968. *Quantum Mechanics.* McGraw Hill, New York.

[384] Schmitt, J. M., and Kumar G., 1998. Optical scattering properties of soft tissue: A discrete particle model. *Appl. Opt.*, 37, 2788–2797.

[385] Schneiderheinze, D. H. P., Hillman, T. R., and Sampson, D. D., 2007. Modified discrete particle model of optical scattering in skin tissue accounting for multiparticle scattering. *Opt. Express*, 15, 15002.

[386] Selamet A., and Arpaci, V. S., 1989. Rayleigh Limit-Penndorf extension. *Int. J. Heat Mass Transfer*, 32, 1809–1820.

[387] Shah, G. A., 1970. Scattering of plane electromagnetic waves by infinite concentric circular cylinders at oblique incidence. *Mon. Not. R. Astron. Soc.*, 148, 93–102.

[388] Shang, Q., Wu,Z, Qu, T, Li , Z., and Gong, L., 2014. Analytical solution of plane wave scattering from a large sized chiral sphere. 10-th Int. conf. on laser-light and interactions with particles, Marseille, France.

[389] Shapovalov, K. A., 2014. *Light Scattering by Nonspherical Particles in the RGD Approximation: Single Scattering.* Lap LAMBERT Academic Publishing, Saarbrücken.

[390] Sharma, S. K., 1989. Approximate formulas for scattering of light by oriented infinitely long homogeneous soft circular cylinders. *J. Mod. Opt.*, 36, 399–404.

[391] Sharma, S. K., 1993. A modified anomalous diffraction approximation for intermediate size soft particles. *Opt. Commun.*, 100, 13–18.

[392] Sharma, S. K., 1994. On the validity of soft particle approximations for the scattering of light by infinitely long homogeneous cylinders. *J. Mod. Opt.*, 41, 827–838.

[393] Sharma, S. K., 2013. A review of approximate analytic light scattering phase functions. *Light Scattering Reviews*, 9, 53–100.

[394] Sharma, S. K., 2015. A modified Rayleigh–Gans–Debye formula for small angle X-ray scattering by interstellar dust grains. *Astrophys. Space Sci.*, 357: 80.

[395] Sharma, S. K., and Banerjee, S., 2003. Role of approximate phase functions in Monte Carlo simulation of light propagation in tissues. *J. Opt. A: Pure Appl. Opt.*, 5, 294–302.

[396] Sharma, S. K., and Banerjee, S., 2005. Volume concentration and size dependence of diffuse reflectance in a fractal tissue model. *Med. Phys.*, 32, 1767–1774.

[397] Sharma, S. K., and Banerjee, S., 2012. On the validity of the anomalous diffraction approximation in light-scattering studies of soft biomedical tissues. *J. Mod. Opt.*, 59, 701–709.

[398] Sharma, S. K., and Jones, A. R., 2000. On the validity of an approximate formula for the scattering and absorption of light by a large sphere with highly absorbing spherical inclusions. *J. Phys. D(GB)*, 33, 584–588.

[399] Sharma, S. K., and Jones, A. R., 2003. Absorption and scattering of electromagnetic radiation by a large absorbing sphere with highly absorbing inclusions. *J. Quant. Spectrosc. Radiat. Transf.*, 79, 1051–1060.

[400] Sharma, S. K., and Roy, A. K., 2000. New approximate phase functions: Test for nonspherical particles. *J. Quant. Spectrosc. Radiat. Transf.*, 64, 327–337.

[401] Sharma, S. K., and Roy, A. K., 2008. On the validity of phase functions used in calculations of electromagnetic wave scattering by interstellar dust. *Astrophys. J. Suppl.*, 177, 546–550.

[402] Sharma, S. K., and Somerford, D. J., 1988. Modified Rayleigh–Gans–Debye approximation applied to sizing transparent homogeneous long fibres of intermediate size. *J. Phys. (UK)*, D21, 403–406.

[403] Sharma, S. K., and Somerford, D. J., 1990. The accuracy of eikonal approximation for sizing particles in cohesive sediments by light scattering. *Water Res.*, 24, 447–450.

[404] Sharma, S. K., and Somerford, D. J., 1992. An analysis of scattering patterns of some hydrological particles. *J. Environ. Sci. Health A*, 27, 153–164.

[405] Sharma, S. K., and Somerford, D. J., 1994. An approximation method for the backward scattering of light by a soft spherical obstacle. *J. Mod. Opt.*, 41, 1433–1444.

[406] Sharma, S. K., and Somerford, D. J., 1996. On the relationship between the S-approximation and Hart–Montroll approximation. *J. Opt. Soc. Am. A*, 13, 1285–1286.

[407] Sharma, S. K., and Somerford, D. J., 1999, Scattering of light in the eikonal approximation. *Progress in Optics*, 39, 211–290.

[408] Sharma S. K., and Somerford, D. J., 2006. *Light Scattering by Optically Soft Particles: Theory and Applications*. Springer/Praxis, UK.

[409] Sharma, S. K., Ghosh, G., and Somerford, D. J., 1997b. The S-approximation for light scattering by an infinitely long cylinder. *Appl. Opt.*, 36, 6109–6114.

[410] Sharma, S. K., Powers, S. R., and Somerford, D. J., 1981. Investigation of domains of validity of approximation methods in light scattering from long cylinders. *Opt. Acta*, 28, 1439–1446.

[411] Sharma, S. K., Roy, A. K., and Somerford, D. J., 1998. New approximate phase function for the scattering of unpolarized light by dielectric particles. *J. Quant. Spectrosc. Radiat. Transf.*, 60, 1001–1010.

[412] Sharma, S. K., Somerford D. J., and Sharma, S., 1982. Investigation of validity domains of corrections to the eikonal approximation in forward light scattering from homogeneous spheres. *Opt. Acta*, 29, 1677–1682.

[413] Sharma, S. K., Somerford, D. J., and Roy, A. K., 1997a. Simple formulae within the framework of anomalous diffraction approximation for light scattered by an infinite long cylinder. *J. Opt.: Pure Appl. Opt. A*, 6, 565–575.

[414] Sharma, S. K., Somerford, D. J., and Shah, H. S., 1999. A new method to calculate Legendre coefficients for use in multiple scattering problems. *J. Opt. (India)*, 123–131.

[415] Shatilov, A. V., 1960. On the scattering of light by dielectric ellipsoids comparable to the light wavelength. I. *Opt. Spectrosc.*, 9, 86–91 [In Russian].

[416] Shen, Y., Draine, B. T., and Johnson, E. T., 2008. Modeling porous dust grains with ballistic aggregates I. Geometry and optical properties. *Astrophys. J.*, 689, 260–275.

[417] Shepelevich, N. V., Prostakova, I. V., and Lopatin, V. N., 1999. Extrema in the light scattering indicatrix of a homogeneous spheroid. *J. Quant. Spectrosc. Radiat. Transf.*, 63, 353–367.

[418] Shepelevich, N. V., Prostakova, I. V., and Lopatin, V. N., 2001. Light scattering by optically soft randomly oriented spheroids. *J. Quant. Spectrosc. Radiat. Transf.*, 70, 375–381.

[419] Sheppard, C. J. R., 2007. Fractal model of light scattering in biological tissue and cells. *Opt. Lett.*, 32, 142–144.

[420] Shifrin K. S., 1952. Light scattering by two layered particles. *Izv. AN USSR Ser. Geophys.*, N2, 15–20.

[421] Shifrin, K. S., and Punina, V. A., 1968. Light scattering indicatrix in the region of small angles. *Bull. Izv. Acad. Sci. USSR Atmos. Oceanic Phys.*, 4, 450–453.

[422] Shifrin, K. S., and Tonna, G. S., 1992. Simple formula for absorption coefficient of weakly refracting particles. *Opt. Spectrosc.*, 72, 487–490.

[423] Shifrin, K. S., and Zolotov, I. S., 1993. Remark about the notation used for calculating the electromagnetic field scattered by a spherical particle. *Appl. Opt.*, 32, 5397–5398.

[424] Shimizu, K., 1983. Modification of the Rayleigh–Debye scattering. *J. Opt. Soc. Am.*, 73, 504–507.

[425] Shipley, S. T., Wienman, J. A., 1978. A numerical study of scattering by large dielectric spheres. *J. Opt. Soc. Am.*, 68, 130–134.

[426] Shore, R. A., 2015. Scattering of an electromagnetic linearly polarized plane wave by a multilayered sphere. *IEEE A P Magazine*, 57, 69–116.

[427] Siewert, C. E., and Williams, M. M. M., 1977. The effect of anisotropic scattering on the critical slab problem in neutron transport theory using a synthetic kernel. *J. Phys. D: Appl. Phys.*, 10, 2031–2040.

[428] Small, A., Hong, S., and Pine, D., 2005. Scattering properties of core-shell particles in plastic matrices. *J. Polym. Sci. Part B.*, 43, 3534–3548.

[429] Smart, C., and Vand, V., 1964. Approximate formula for the total scattering of electromagnetic radiation by spheres. *J. Opt. Soc. Am.*, 54, 1232–1234.

[430] Smith, R. K., and Dwek, E., 1998. Soft X-ray scattering and halos from dust. *Astrophys. J.*, 503, 831–842.

[431] Sokoletsky, L. G. et al., 2009. A comparison of analytical radiative transfer solutions of plane albedo for natural waters. *J. Quant. Spectrosc. Radiat. Transf.*, 110, 1132–1146.

[432] Sokolov, A., Chami, M., Dimitriev, E., and Khomenko, G., 2010. Parameterization of volume scattering function of coastal waters based on the statistical approach. *Opt. Express*, 18, 4615–4636.

[433] Sorensen, C. M., 2013. Q space analysis of scattering by particles: A review. *J. Quant. Spectrosc. Radiat. Transf.*, 131, 3–12.

[434] Steiner, B., Berge, B., Gausmann, R., Rohmann, J., and Rühl, E., 1999, Fast *in situ* sizing technique for single levitated aerosols. *Appl. Opt.*, 38, 1523–1529.

[435] Stephens, G. L., 1984. Scattering of plane waves by soft obstacles: Anomalous diffraction theory for circular cylinders. *Appl. Opt.*, 23, 954–959.

[436] Stroem, S., 1974. Scattering from an arbitrary number of scatterers with continuously varying electromagnetic properties. *Phys. Rev. D.*, 10, 2685–2690.

[437] Stratton, J. A., 1941. *Electromagnetic theory.* McGraw-Hill Book Co., New York.

[438] Streekstra, G. J., 1994. The deformation of red blood cells in a couette flow. Ph.D thesis, University of Utrecht, The Netherlands.

[439] Streekstra, G. J., Hoekstra, A. G., Evert-Jan, N, and Heethar, M., 1993. Light scattering by red blood cells in ektacytometry: Fraunhofer versus anomalous diffraction. *Appl. Opt.*, 32, 2266–2272.

[440] St*ü*binger, T., K*ü*hler, U., and Witt, W., 2010. Verification of Mie scattering algorithms by extreme precision calculations. *WCPT*, 6, 1–4.

[441] Sudiarta, I. W. and Chylek, P., 2001a. Mie scattering formalism for spherical particles embedded in an absorbing medium. *J. Opt. Soc. Am. A*, 18, 1275–1278.

[442] Sudiarta, I. W. and Chylek, P., 2001b. Mie extinction efficiency of a large spherical particle embedded in an absorbing medium. *J. Quant. Spectrosc. Radiat. Transf.*, 70, 709–714.

[443] Sudiarta, I. W. and Chylek, P., 2002. Mie scattering by a spherical particle in an absorbing medium. *Appl. Opt.*, 41, 3545–3546.

[444] Sun, W., and Fu, Q., 1999. Anomalous diffraction theory for randomly oriented hexagonal crystals. *J. Quant. Spectrosc. Radiat. Transf.*, 63, 727–737.

[445] Sun, W., and Fu, Q., 2001. Anomalous diffraction theory for randomly oriented nonspherical particles. *J. Quant. Spectrosc. Radiat. Transf.*, 63, 737–747.

[446] Swanson, N. L., Billard, B. D., and Gennaro, T. L., 1999. Limits of optical transmission measurements with application to particle sizing techniques. *Appl. Opt.* , 38, 5887–5893.

[447] Tien, C. L., Doornink, D. G., and Rafferty, D. A., 1972. Attenuation of visible radiation by carbon smokes. *Combust. Sci. Tech.*, 6, 55–59.

[448] Toublanc, D., 1996. Henyey–Greenstein and Mie phase functions in Monte Carlo radiative transfer computations, *Appl. Opt.*, 35, 3270–3274.

[449] Tuchin, V. V., 1997, Light scattering study of tissues. UFN, 5, 517–539.

[450] Tuomi, T. J., 1980. Light scattering by aerosols with layered humidity-dependent structure. *J. Aerosol Sci.*, 11, 367–375.

[451] Turcu, I., Pop, C. V. L., and Neamtu, S., 2006. High-resolution angle resolved measurement of light scattered at small angles by red blood cells in suspension. *Appl. Opt.*, 45, 1964–1971.

[452] Turcu, I., and Bratfalean, R., 2008a. Narrowly peaked forward light scattering on particulate media I. Assessment of the multiple scattering contributions to the effective phase function. *J. Opt. A: Pure Appl. Opt.*, 10, 015002.

[453] Turcu, I., Bratfalean, R., and Neamtu, S., 2008b. Narrowly peaked forward light scattering on particulate media II. Angular spreading of light scattered by polysterene microspheres. *J. Opt. A: Pure Appl. Opt.*, 10, 075007.

[454] Turcu, I., and Kirillin, M., 2009. Qiasi-ballistic light scattering-analytical model versus Monte Carlo simulations. *J. Phys. Conference Ser..*, 182, 012035.

[455] Turner, L., 1976. Light scattering by ensembles randomly oriented anisotropic particles. *Appl. Opt.*, 15, 1085–1090.

[456] Turzhitsky, V., Radosevich, A., Rogers, J. D., Taflove, A., and Beckman, A., 2010. A predictive model of backscattering at subdiffusion length scales. *Biomed. Opt. Exppress*, 1, 1034–1046.

[457] Twersky, V., 1975. Transparency of pair correlated, random distribution of small scatterers, with application to cornea. *J. Opt. Soc. Am.*, 65, 524–530.

[458] Twomey, S., and Bohren, C. F., 1980. Simple approximations for calculations of absorption in clouds. *J. Atmos. Sci.*, 37, 2086–2094.

[459] van de Hulst, H. C., 1957. *Light Scattering by Small Particles.* John Wiley & Sons, New York.

[460] van der Pol, B., and Bremmer, H., 1937a. The diffraction of electromagnetic waves from an electrical point source round a finitely conducting sphere with applications to radiotelegraphy and the theory of the rainbow. Part 1. *Philos. Mag.*, 24, 141–176.

[461] van der Pol, B., and Bremmer, H., 1937b. The diffraction of electromagnetic waves from an electrical point source round a finitely conducting sphere with applications to radiotelegraphy and the theory of the rainbow. Part 2. *Philos. Mag.*, 24, 825–864.

[462] van der Zee, P., Essenpreis, M., and Delpy, D. T., 1993. Optical properties of brain tissues. *Proc. SPIE*, 1888, 454–465.

[463] Vaudelle, F., 2017, Approximate analytical effective phase function obtained for a thin slab geometry. *J. Quant. Spectrosc. Radiat. Transf.*, 193, 47–56.

[464] Videen, G., and Bickel, W. S., 1992. Light scattering resonances in small spheres. *Phys. Rev. A*, 45, 6008–6012.

[465] Videen, G., and Sun, W., 2003. Yet another look at light scattering from particles in absorbing media. *Appl. Opt.* , 33, 6724–6727.

[466] Videen, G., Ngo, D., Chylek, P., and Pinnick, R. G., 1995. Light scattering from a sphere with an irregular inclusion. *J. Opt. Soc. Am. A*, 12, 922–928.

[467] Vo-Dinh, T. (Ed.), 2015. *Biomedical Photonics Handbook*, 2nd Ed. CRC Press, Boca Raton.

[468] Volkov, N. G., and Kovach, Y. Y., 1990. Scattering of light by inhomogeneous spherical symmetric aerosol particles. *Izvestia Atmos. Ocean Phys.*, 26, 381–385.

[469] Voshchinnikov, N. V., and Mathis, J. S., 1999. Calculating cross-sections of cosmic interstellar grains. *Astrophys. J.*, 526, 257–264.

[470] Voshchinnikov, N. V., and Farfonov, V. G., 2000. Applicability of quasistatic and Rayleigh approximations for spheroidal particles. *Opt. Spectrosc.*, 88, 71–75.

[471] Voshchinnikov, N. V., Videen. G., and Henning. T., 2007. Effective medium theories for irregular fluffy structures: Aggregation of small particles. *App. Opt.*, 46, 4065–4072.

[472] Wait, J. R., 1955. Scattering of a plane wave from a circular dielectric cylinder at oblique incidence. *Can. J. Phys.*, 33, 189–195.

[473] Wait, J. R., 1963. Electromagnetic scattering from a radially inhomogeneous sphere. *Appl. Sci. Res. B*, 10, 441–550.

[474] Wallace, S. J., 1973. Eikonal expansion. *Ann. Phys.*, 78, 190–257.

[475] Walstra, P., 1964. Approximation formulae for light scattering coefficient of dielectric spheres. *Br. J. Appl. Phys.*, 15, 1545–1551.

[476] Wang, R. K., 2000. Modelling optical properties of soft tissues by fractal distribution of scatterers. *J. Mod. Opt.*, 47, 103–120.

[477] Wang, L., Sun, X., and Li, F., 2012. Generalized eikonal approximation for fast retrieval of particle size distribution in spectral extinction technique. *Appl. Opt.*, 51, 2997-3005.

[478] Wax, A., and Backman, V., 2010. *Biomedical Applications of Light Scattering*. McGraw Hill, New York.

[479] Wentzel, G., 1926. Eine verallgemeinerung der quantenbedingungen für die Zwecke der Wellenmechanik. *Z. Phys.*, 38, 518–529.

[480] Whittet, D. C. B., 2003. *Dust in Galactic Environment*. IOP Publishing Limited, Bristol.

[481] Wickramsinghe, N. C., 1973. *Light Scattering Functions for Small Particles with Applications in Astronomy.* Hilger, London.

[482] Wienman, J. A., and Kim, M,-J., 2007. A simple model of the millimeter-wave scattering parameters of randomly oriented aggregates of finite cylindrical ice hydrometeors. *J. Atmos. Sci.*, 64, 634–644.

[483] Windt, W. I., Cash, W., and Kahn, S. M., 2000. The scattering of X-rays by interstellar dust on the microarcsecond scale. *Astrophys. J.*, 528, 306–309.

[484] Wiscombe, W. J., 1977. The delta-M method: Rapid yet accurate radiative flux calculations for strongly asymmetric phase functions. *J. Atmos. Sci.*, 34, 1408–1422.

[485] Wiscombe, W. J., 1980. Improved Mie scattering algorithms. *Appl. Opt.*, 19, 1505–1509.

[486] Wiscombe, W. J., and Chylek, P., 1977. Mie scattering between any two angles. *J. Opt. Soc. Am.*, 67, 572–573.

[487] Wong, K. L., and Chen, H. T., 1992. Electromagnetic scattering by a uniaxial anisotropic sphere. *IEEE Proceedings-H.*, 139, 314–318.

[488] Wriedt, T., 1998. A review of elastic light scattering theories. *Part Part Syst. Chart.*, 15, 67–74.

[489] Wriedt, T., and Comeberg, U., 1998. Comparison of computational scattering methods. *J. Quant. Spectrosc. Radiat. Transfer*, 60, 411–423.

[490] Wu, Z-S., Shang, Q-C., and Li, Z-J., 2012. Calculation of electromagnetic scattering by a large chiral sphere. *Appl. Opt.*, 51, 6661–6668.

[491] Wyatt, P. J., 1962. Scattering of electromagnetic plane waves from inhomogeneous spherically symmetric objects. *Phys. Rev.*, 127, 1837–1842.

[492] Xie, H-Y., Ng, M-Y., and Chang, Y-C., 2010. Analytic solutions to light scattering by plasmonic nanoparticles with nearly spherical shapes and nonlocal effects. *J. Opt. Soc. Am. A*, 27, 2411–2422.

[493] Xu, F., Lock, J. A., and Tropea, C., 2010a. Debye series for light scattering by a spheroid. *J. Opt. Soc. Am. A*, 27, 671–686.

[494] Xu, F., Lock, J. A., and Gouesbet, G., 2010b. Debye series for light scattering by a nonspherical particle. *Phys. Rev A*, 81, 043824.

[495] Xu, L., Chen, Z., Taflove, A., and Backman, V., 2004. Equiphase sphere approximation for analysis of light scattering by arbitrarily shaped nonspherical particles. *Appl. Opt.*, 43, 4497–4505.

[496] Xu, M., 2003. Light extinction and absorption by arbitrarily oriented finite circular cylinders by use of geometrical path statistics. *Appl. Opt.*, 42, 6710–6723.

[497] Xu, M., 2016. Diagnosis of phase function of random media from light reflectance. *Scientific Reports*, 6, 22535.

[498] Xu, M., and Alfano, R. R., 2005. Fractal mechanism of light scattering in biological tissue and cells. *Opt. Lett.*, 30, 3051–3053.

[499] Xu, M., and Katz, A., 2008. Statistical interpretation of light anomalous diffraction by small particles and its applications in bio-agent detection and monitoring. *Light Scattering Reviews*, 3, 27–63.

[500] Xu, M., Lax, M., and Alfano, R. R., 2003. Anomalous diffraction of light with geometrical path statistics of rays and a Gaussian ray approximation. *Opt. Lett.*, 28, 179–181.

[501] Xu, R., 2000. *Particle Characterization: Light Scattering Methods.* Springer, Netherlands.

[502] Yang, P., Gao, B., Wiscombe, W. J., Mishchenko, M. I., Platnick, S. E., Huang, H., Baum, B., Hu, Y. X., Winker, D. M., Tsay, S., and Park, S. K., 2002. Inherent and apparent scattering properties of coated or uncoated spheres in an absorbing host medium. *Appl. Opt.*, 41, 2740–2759.

[503] Yang, P., Zhang, Z., Baum, B. A., Huang, H. L., and Hu, Y., 2004. A new look at the anomalous diffraction theory (ADT): Algorithm in cumulative projected-area distribution. *J. Quant. Spectrosc. Radiat. Transf.*, 89, 421–442.

[504] Yaroslavsky, A. N., I. V. Yaroslavsky, T. Goldbach, and H.-J. Schwarzmaier, 1997. Different phase function approximations to determine optical properties of blood: A comparison. *SPIE* , 2982, 324–330.

[505] Yaroslavsky, A. N., I. V. Yaroslavsky, T. Goldbach, and H.-J. Schwarzmaier, 1999. Influence of the scattering phase function approximation on the optical properties of blood determined from the integrating sphere measurements. *J. Biomed. Opt.*, 4, 47–53.

[506] Yeh, C., and Lindgreen, G., 1977. Computing the propagation characteristics of radially stratified fibres: An efficient method. *Appl. Opt.*, 16, 483–493.

[507] Yin, J., and Pilon, L., 2006. Efficiency factors and radiation characteristics of spherical scatterers in an absorbing medium. *J. Opt. Soc. Am. A*, 23, 2784–2796.

[508] Yoon, G., 1988. Absorption and scattering of laser light in biological media-mathematical modelling and methods for determining optical properties. Ph.D thesis. University of Texas, Austin.

[509] You-wei, H., et al., 2012. Comparison study of several underwater light scattering phase functions. *Optoelectronic Lett.*, 8, 0233–0236.

[510] Yurkin, M. A., and Hoekstra, A. G., 2011. The discrete dipole approximation code ADDA: Capabilities and known limitations. *J. Quant. Spectrosc. Radiat. Transf.*, 112, 2234–2247.

[511] Zege, E. P., and Kokhanovsky, A., 1989. Approximation of the anomalous diffraction to coated spheres. *Izv. Atmos. Oceanic Phys.*, 25, 1195–1201.

[512] Zhao, J., Shi, G., Che, H., and Cheng, G., 2006. Approximation of the scattered phase functions of particles. *Adv. Atmos. Sci.*, 23, 802–808.

[513] Zhao, Z., and Sun, X., 2010. Error analysis of using Henyey–Greenstein phase function in Monte Carlo radiative transfer simulations. *Proceedings in Progress in Electromagnetic Research Symposium*, Xi'an, China.

[514] Zhou, Y., He., Q. S., and Zheng, X. J., 2005. Attenuation of electromagnetic wave propagation in sandstorms incorporating charged sand particles. *Eur. Phys. J. E.*, 17, 181–187.

[515] Zolek, N. S., Wojtriewicz, S., and Liebert A., 2008. Correction of anisotropy coefficient in original Henyey–Greenstein phase function for Monte Carlo simulation of light transport in tissues. *Biocybernetics Biomed. Eng.*, 28, 59–73.

[516] Zude, M., 2009. *Optical Monitoring of Fresh and Processed Agricultural Crops*. CRC Press, Boca Raton.

Index